Unstoppable Owl

Unstoppable Owl

Name: Date: Score:

1) Double-Digit Addition (Without Regrouping)

Before learning to add 2-digit numbers, it's helpful to master basic addition facts. If you're not comfortable adding single-digit numbers, you may want to consider using the book to the right first.

The addition problems in this section do not require regrouping–which makes them some of the easiest problems for learning to add 2-digit numbers. All you have to do is add the digits, one column at a time.

Step 1: Add the digits in the ones column (the column to the right.) Write the answer directly below.
Step 2: Add the digits in the tens column and write that answer directly below.

Example:

	5	3
+	2	6

53 + 26 =

Step 1:

	5	3
+	2	6
		9

3 + 6 = **9**

Step 2:

	5	3
+	2	6
	7	9

5 + 2 = **7**

Answer:

	5	3
+	2	6
	7	9

53 + 26 = **79**

Directions: *Solve each addition problem.*
Remember to add the digits in each column, one column at a time.

1)

	4	3
+	2	5

2)

	8	2
+	1	3

3)

	1	4
+	6	5

4)

	3	0
+	3	7

5)

	3	6
+	4	2

6)

	1	5
+	1	1

7)

	2	2
+	1	3

8)

	4	2
+	4	7

Name: Date: Score: *Unstoppable Owl*

2) Double-Digit Addition (Without Regrouping)

1) 21 + 53 =

2) 38 + 40 =

3) 16 + 42 =

4) 72 + 22 =

5) 60 + 15 =

6) 38 + 31 =

7) 12 + 16 =

8) 26 + 60 =

9) 25 + 21 =

10) 46 + 33 =

11) 17 + 51 =

12) 23 + 23 =

13) 13 + 42 =

14) 41 + 44 =

15) 52 + 15 =

16) 21 + 30 =

17) 31 + 28 =

18) 50 + 20 =

19) 23 + 16 =

20) 40 + 13 =

21) 74 + 15 =

22) 31 + 41 =

23) 44 + 22 =

24) 80 + 19 =

Name: Date: Score: Unstoppable Owl

3) **Double-Digit Addition** (Without Regrouping)

You've now had practice adding two double-digit numbers. But how are 1-digt and 2-digit numbers added? Well, they are added the same way. Add one column at a time. When there is only one number in a column, it is the same as adding that number with 0 (adding it with nothing.)

Step 1: Add the digits in the ones column.
Step 2: Add the digits in the tens column.

Example:

		7
+	6	1

7 + 61 =

Step 1:

		7
+	6	1
		8

↑

7 + 1 = **8**

Step 2:

		7
+	6	1
	6	8

↑

0 + 6 = **6**

Answer:

		7
+	6	1
	6	8

7 + 61 = **68**

Directions: *Solve each addition problem. Add the digits in each column, one column at a time. When there is only one number in a column, it is the same as adding that number with 0.*

1)

		5
+	2	3

2)

	8	0
+		6

3)

	7	7
+		2

4)

		3
+	4	1

5)

	9	4
+		3

6)

		5
+	6	2

7)

	1	3
+		3

8)

		6
+	2	2

9)

		7
+	5	1

10)

		4
+	1	4

11)

	3	0
+		8

12)

	4	4
+		5

Unstoppable Owl

Name: Date: Score:

4) **Double-Digit Addition** (Without Regrouping)

1) 42 + 35 =

2) 4 + 81 =

3) 36 + 32 =

4) 13 + 35 =

5) 27 + 21 =

6) 43 + 12 =

7) 32 + 54 =

8) 93 + 4 =

9) 14 + 34 =

10) 50 + 17 =

11) 31 + 51 =

12) 23 + 73 =

13) 55 + 2 =

14) 40 + 20 =

15) 14 + 65 =

16) 24 + 21 =

17) 36 + 32 =

18) 24 + 53 =

19) 43 + 46 =

20) 12 + 21 =

21) 15 + 83 =

22) 6 + 41 =

23) 20 + 20 =

24) 33 + 52 =

Unstoppable Owl

Name: Date: Score:

5) Double-Digit Addition (Regrouping the Ones Column)

Regrouping is needed when the numbers added in one column are greater than or equal to 10. For example, if you add the numbers in the ones column and they equal 13, 13 cannot be written in the ones column below the problem. The 13 represents 1 ten and 3 ones. The 3 ones can remain in the ones column, but the 1 ten needs to be regrouped to the tens column. People often do this by writing the 3 in the ones column below the problem and writing the 1 above the tens column to be added with the other tens in that column.

See the 1 written above the tens column below? It was regrouped from the ones column to the tens column. Then it was added with the 5 and the 2 that were already in the tens column.

Step 1: Add the digits in the ones column.
Step 2: Regroup from the ones column to the tens column (*if necessary*).
Step 3: Add the digits in the tens column.

Step 1:

	5	7
+	2	6
		3

↑

7 + 6 = **13**

Step 2:

	1 ←	
	5	7
+	2	6
		3

Regroup

Step 3:

	1	
	5	7
+	2	6
	8	3

↑

1 + 5 + 2 = **8**

Answer:

	1	
	5	7
+	2	6
	8	3

57 + 26 = **83**

Directions: *Solve each addition problem.*
Many problems will require regrouping, but some may not.

1)

	3	4
+	2	8

2)

	5	7
+	3	7

3)

	2	3
+	4	6

4)

	1	6
+	4	9

5)

	1	8
+	2	7

6)

	3	9
+	4	1

7)

	2	6
+	6	5

8)

	1	4
+	7	4

Name: Date: Score: *Unstoppable Owl*

6) **Double-Digit Addition** (Regrouping the Tens Column)

The ones column is not the only column that may require regrouping. When adding 2-digit numbers, the tens column may require regrouping too. When regrouping the tens column, part of the value is regrouped from the tens column to the hundreds column. This is because 10 tens are equal to 1 hundred. That 1 hundred needs to be moved to the hundreds column.

Step 1: Add the digits in the ones column.
Step 2: Regroup from the ones column to the ten's column (*if necessary*).
Step 3: Add the digits in the tens column.
Step 4: Regroup from the ones column to the tens column (*if necessary*).
Step 5: Add the digits in the hundreds column (*if there are any*).

Steps 1 and 2:

	1 ←	
	6	7
+	8	9
		6

7 + 9 = **16**
Regroup

Steps 3 and 4:

1 ↗	1	
	6	7
+	8	9
	5	6

1 + 6 + 8 = **15**
Regroup

Step 5:

1	1	
	6	7
+	8	9
1	5	6

1 + 0 = **1**

Answer:

1	1	
	6	7
+	8	9
1	5	6

67 + 89 = **156**

Directions: *Solve each addition problem.*
Many problems will require regrouping in one or two columns, but regrouping may not be necessary.

1)

	8	5
+	3	9

2)

	5	1
+	5	7

3)

	9	6
+	8	4

4)

	4	6
+	5	7

5)

	2	3
+	7	4

6)

	1	5
+	7	8

7)

	6	5
+	4	2

8)

	8	2
+	8	9

Name: Date: Score: Unstoppable Owl

7) Double-Digit Addition

1)

	6	5
+	6	3

2)

	8	3
+	1	9

3)

	4	8
+	3	5

4)

	5	3
+	7	8

5)

	6	2
+	4	7

6)

	1	5
+	9	5

7)

	4	9
+	2	8

8)

	9	7
+	9	7

9)

	8	7
+	5	4

10)

	3	2
+	8	0

11)

	3	4
+	5	6

12)

	2	6
+	7	5

13)

	5	6
+	9	6

14)

	4	5
+	4	5

15)

	4	6
+	7	3

16)

	1	7
+	9	8

17)

	3	6
+	2	9

18)

	2	5
+	8	3

19)

	7	9
+	6	4

20)

	5	6
+	1	2

21)

	8	9
+	6	7

22)

	4	3
+	5	8

23)

	2	2
+	8	4

24)

	5	6
+	6	5

Name: Date: Score: Unstoppable Owl

8) Double-Digit Addition

1) 42 + 86 = ___

2) 47 + 58 = ___

3) 32 + 49 = ___

4) 75 + 66 = ___

5) 60 + 85 = ___

6) 59 + 59 = ___

7) 35 + 74 = ___

8) 46 + 94 = ___

9) 48 + 55 = ___

10) 38 + 39 = ___

11) 54 + 73 = ___

12) 87 + 31 = ___

13) 50 + 56 = ___

14) 74 + 79 = ___

15) 63 + 57 = ___

16) 99 + 23 = ___

17) 84 + 37 = ___

18) 43 + 45 = ___

19) 67 + 95 = ___

20) 65 + 15 = ___

21) 79 + 89 = ___

22) 64 + 36 = ___

23) 85 + 85 = ___

24) 78 + 46 = ___

Name: Date: Score: Unstoppable Owl

9) Double-Digit Addition

1)	2)	3)	4)
46 + 77	54 + 39	73 + 86	57 + 59

5)	6)	7)	8)
78 + 28	34 + 46	83 + 43	68 + 75

9)	10)	11)	12)
60 + 50	47 + 98	65 + 66	76 + 95

13)	14)	15)	16)
75 + 97	58 + 63	76 + 46	52 + 98

17)	18)	19)	20)
49 + 39	57 + 46	87 + 24	51 + 83

21)	22)	23)	24)
36 + 75	95 + 78	73 + 29	86 + 64

Name: Date: Score: *Unstoppable Owl*

10) Double-Digit Addition

1) 65 + 78 = ___

2) 98 + 18 = ___

3) 67 + 27 = ___

4) 56 + 73 = ___

5) 42 + 56 = ___

6) 75 + 93 = ___

7) 85 + 36 = ___

8) 24 + 84 = ___

9) 99 + 99 = ___

10) 59 + 81 = ___

11) 46 + 76 = ___

12) 35 + 97 = ___

13) 36 + 28 = ___

14) 57 + 59 = ___

15) 63 + 65 = ___

16) 92 + 49 = ___

17) 45 + 48 = ___

18) 93 + 87 = ___

19) 28 + 97 = ___

20) 64 + 56 = ___

21) 37 + 85 = ___

22) 56 + 26 = ___

23) 47 + 32 = ___

24) 69 + 95 = ___

Name: Date: Score: Unstoppable Owl

11) Double-Digit Addition (Without Guidelines)

So far, the problems in this book have had guidelines to help you learn and keep your calculations organized. The goal is for you to be able to write the problems yourself, and then solve them. Writing the problems is not difficult. Just be sure that you align the columns properly. This means the digits in the ones column should be positioned directly above one another. The same needs to be true for the digits in the tens column.

Example 1: 23 + 45 =		**Example 2:** 6 + 31 =	
Columns Aligned	Columns Not Aligned	Columns Aligned	Columns Not Aligned
23 +45 68	23 +45 473	6 +31 37	6 +31 91
23 + 45 = **68**	23 + 45 = **473** ???	6 + 31 = **37**	6 + 31 = **91** ???
Correct Answer	**Wrong Answer**	**Correct Answer**	**Wrong Answer**

Directions: *Solve each addition problem.*
Many problems will require regrouping in one or two columns, but regrouping may not be necessary.

1) **27 + 56 =**

2) **83 + 9 =**

3) **54 + 63 =**

4) **93 + 77 =**

5) **5 + 36 =**

6) **64 + 73 =**

7) **55 + 59 =**

8) **28 + 68 =**

Name: Date: Score: Unstoppable Owl

12) Double-Digit Addition

1) **47 + 25 =**

2) **83 + 58 =**

3) **94 + 97 =**

4) **62 + 89 =**

5) **36 + 45 =**

6) **74 + 66 =**

7) **85 + 27 =**

8) **63 + 7 =**

9) **8 + 97 =**

10) **29 + 25 =**

11) **67 + 84 =**

12) **53 + 39 =**

13) **45 + 45 =**

14) **76 + 95 =**

15) **38 + 78 =**

16) **43 + 24 =**

17) **86 + 85 =**

18) **6 + 79 =**

19) **87 + 53 =**

20) **74 + 28 =**

Name: Date: Score: Unstopppable Owl

13) Double-Digit Addition

1) **28 + 94 =**

2) **45 + 8 =**

3) **27 + 36 =**

4) **55 + 93 =**

5) **42 + 59 =**

6) **74 + 88 =**

7) **57 + 44 =**

8) **83 + 65 =**

9) **38 + 39 =**

10) **57 + 62 =**

11) **49 + 59 =**

12) **84 + 99 =**

13) **49 + 86 =**

14) **67 + 85 =**

15) **94 + 56 =**

16) **57 + 31 =**

17) **97 + 75 =**

18) **9 + 76 =**

19) **15 + 65 =**

20) **47 + 87 =**

Name: Date: Score: Unstoppable Owl

14) Double-Digit Addition

1) **64 + 88 =**

2) **94 + 78 =**

3) **49 + 67 =**

4) **57 + 83 =**

5) **61 + 87 =**

6) **7 + 65 =**

7) **51 + 79 =**

8) **26 + 8 =**

9) **76 + 85 =**

10) **94 + 98 =**

11) **63 + 28 =**

12) **17 + 16 =**

13) **5 + 47 =**

14) **38 + 43 =**

15) **52 + 79 =**

16) **35 + 72 =**

17) **89 + 4 =**

18) **13 + 68 =**

19) **96 + 74 =**

20) **63 + 58 =**

Name: Date: Score: *Unstoppable Owl*

15) Triple-Digit Addition

Adding 3-digit numbers is similar to adding 2-digit numbers. There's just an extra column to add—the hundreds column. Sometimes, you may need to regroup from the hundreds column to the thousands column. But the process is the same as regrouping any other column.

Step 1: Add the digits in the ones column. Regroup *if necessary*.
Step 2: Add the digits in the tens column. Regroup *if necessary*.
Step 3: Add the digits in the hundreds column. Regroup *if necessary*.

Example 1: 548 + 726 =

1		1	
	5	4	8
+	7	2	6
1,	2	7	4

548 + 726 = **1,274**

Example 2: 370 + 295 =

	1		
	3	7	0
+	2	9	5
	6	6	5

370 + 295 = **665**

Example 3: 657 + 743 =

1	1	1	
	6	5	7
+	7	4	3
1,	4	0	0

657 + 743 = **1,400**

Directions: *Solve each addition problem.*
Many problems will require regrouping in one or more columns, but regrouping may not be necessary.

1)

	5	0	6
+	7	2	6

2)

	8	5	1
+	6	9	8

3)

	4	8	9
+	3	5	7

4)

	6	8	5
+	9	1	5

5)

	3	2	4
+	5	7	4

6)

	7	0	8
+	7	9	4

7)

	5	7	8
+	6	4	9

8)

		6	5
+	7	7	4

9)

	9	5	8
+		7	1

Name: Date: Score: Unstoppable Owl

16) Triple-Digit Addition

1)

	2	8	7
+	2	6	3

2)

	1	7	5
+	5	8	0

3)

	6	2	8
+	7	9	2

4)

	7	5	6
+	9	2	5

5)

		4	3
+	8	7	9

6)

	1	9	3
+	1	4	3

7)

	6	8	4
+	3	2	9

8)

	5	3	9
+	8	7	9

9)

	6	3	7
+	5	6	8

10)

			8
+	9	9	3

11)

	2	8	4
+	4	5	3

12)

	7	5	8
+	7	5	6

13)

	6	5	3
+	4	9	2

14)

	7	0	5
+	3	0	5

15)

	9	9	9
+	9	9	9

16)

	6	8	1
+	1	1	9

17)

	4	6	8
+	5	2	7

18)

	6	4	5
+	6	8	7

Name: Date: Score:

Unstoppable Owl

17) Triple-Digit Addition

1)

	5	1	8
+	9	3	6

2)

	4	5	6
+	7	4	6

3)

	7	2	5
+	5	6	1

4)

		9	4
+	8	2	5

5)

	4	8	7
+	9	5	3

6)

	9	2	5
+		8	2

7)

	6	3	4
+	5	8	9

8)

	1	4	8
+	4	2	6

9)

	2	6	8
+	7	5	3

10)

	3	9	4
+			8

11)

	7	5	9
+	7	4	3

12)

		7	5
+	7	5	0

13)

	6	2	7
+	4	5	8

14)

	9	4	2
+	9	7	6

15)

	6	1	8
+	8	9	5

16)

	3	9	5
+	3	6	7

17)

	6	3	5
+	8	7	5

18)

	9	5	7
+		6	3

Name: Date: Score:

Unstoppable Owl

18) Triple-Digit Addition

1)

	2	9	3
+	1	4	6

2)

	4	4	8
+	6	8	2

3)

	7	6	2
+	5	3	9

4)

	6	8	4
+		5	3

5)

	2	8	4
+	3	8	6

6)

	8	3	6
+	8	4	7

7)

	9	0	6
+	3	9	7

8)

	7	4	5
+	3	8	5

9)

	6	7	8
+	5	7	2

10)

	5	7	6
+	4	9	5

11)

	1	5	3
+	4	7	8

12)

	4	7	0
+	5	4	0

13)

	3	6	5
+	9	6	8

14)

	4	5	7
+	3	1	2

15)

			6
+	5	8	7

16)

	7	9	4
+	4	9	6

17)

	3	8	9
+		7	4

18)

	5	6	8
+	5	1	9

Name: Date: Score: Unstoppable Owl

19) Triple-Digit Addition

1)

	7	3	6
+	2	8	5

2)

	4	4	3
+	8	9	6

3)

	2	4	8
+	9	5	3

4)

	7	7	7
+	7	7	7

5)

	5	4	3
+	9	7	8

6)

	4	5	6
+	7	9	4

7)

	3	8	6
+	4	7	9

8)

	4	3	6
+	8	2	5

9)

		9	5
+	9	2	3

10)

	2	6	9
+	7	5	2

11)

	4	0	6
+	9	3	7

12)

	3	7	4
+	5	8	6

13)

	5	4	9
+	6	8	2

14)

	4	7	9
+	5	6	8

15)

	2	6	5
+	2	4	3

16)

	7	6	5
+	2	9	5

17)

	4	8	5
+	4	1	6

18)

	8	6	7
+	8	1	9

Name: Date: Score:

20) Triple-Digit Addition (Without Guidelines)

You've had a chance to practice adding triple-digit numbers using guidelines. Now it's time to write the problems yourself. Remember to align the columns and place values.

(Refer to page 11 if you would like to review these skills.)

1) **983 + 574 =**

2) **284 + 726 =**

3) **519 + 487 =**

4) **67 + 158 =**

5) **792 + 428 =**

6) **375 + 826 =**

7) **749 + 975 =**

8) **324 + 365 =**

9) **948 + 57 =**

10) **288 + 76 =**

11) **957 + 973 =**

12) **638 + 592 =**

Name: Date: Score: Unstoppable Owl

21) Triple-Digit Addition

1) **568 + 276 =**

2) **283 + 485 =**

3) **947 + 865 =**

4) **479 + 532 =**

5) **86 + 264 =**

6) **528 + 674 =**

7) **974 + 58 =**

8) **475 + 293 =**

9) **846 + 795 =**

10) **372 + 584 =**

11) **189 + 245 =**

12) **473 + 641 =**

13) **573 + 457 =**

14) **907 + 905 =**

15) **724 + 618 =**

Name: Date: Score: *Unstoppable Owl*

22) Triple-Digit Addition

1) **738 + 562 =**

2) **164 + 375 =**

3) **283 + 957 =**

4) **274 + 268 =**

5) **562 + 319 =**

6) **716 + 285 =**

7) **736 + 485 =**

8) **462 + 379 =**

9) **89 + 234 =**

10) **581 + 547 =**

11) **165 + 193 =**

12) **275 + 936 =**

13) **507 + 268 =**

14) **794 + 485 =**

15) **359 + 78 =**

Name: Date: Score: *Unstoppable Owl*

23) Triple-Digit Addition

1) **460 + 392 =**

2) **174 + 586 =**

3) **371 + 425 =**

4) **984 + 897 =**

5) **683 + 415 =**

6) **527 + 674 =**

7) **844 + 776 =**

8) **64 + 754 =**

9) **881 + 695 =**

10) **227 + 948 =**

11) **318 + 850 =**

12) **273 + 494 =**

13) **526 + 574 =**

14) **867 + 946 =**

15) **558 + 972 =**

Name: Date: Score: Unstoppable Owl

24) Triple-Digit Addition

1) **619 + 733 =**

2) **621 + 184 =**

3) **589 + 72 =**

4) **175 + 435 =**

5) **766 + 928 =**

6) **409 + 637 =**

7) **67 + 934 =**

8) **429 + 615 =**

9) **984 + 942 =**

10) **255 + 255 =**

11) **783 + 8 =**

12) **460 + 752 =**

13) **849 + 576 =**

14) **317 + 328 =**

15) **745 + 959 =**

Name: Date: Score: Unstoppable Owl

25) Adding Large Numbers

You've added 2-digit numbers. You've added 3-digit numbers. The steps to add larger numbers are the same. Focus on adding the digits, one column at a time, until all the columns have been added. Sometimes, you may need to regroup. Sometimes, regrouping will not be necessary.

Example 1: 7,876 + 5,514 =

1	1		1	
	7,	8	7	6
+	5,	5	1	4
1	3,	3	9	0

7,876 + 5,514 = **13,390**

Example 2: 648,640 + 803,978 =

1		1	1	1		
	6	4	8,	6	4	0
+	8	0	3,	9	7	8
1,	4	5	2,	6	1	8

648,640 + 803,978 = **1,452,618**

Directions: *Solve each addition problem.*
Many problems will require regrouping in one or more columns, but regrouping may not be necessary.

1)

	9,	4	0	5
+	2,	8	6	5

2)

	7,	5	8	6
+	6,	3	8	5

3)

	8	4,	5	2	6
+		7,	6	4	1

4)

	9	4,	8	7	5
+	9	1,	5	4	9

5)

	6	5	8,	3	2	7
+	3	7	8,	0	9	7

6)

	4	8	3,	9	8	8
+	7	5	8,	3	1	5

Name: Date: Score:

Unstoppable Owl

26) Adding Large Numbers

1)

	2,	6	8	3
+	5,	1	7	6

2)

	7,	5	4	5
+	4,	9	8	7

3)

		8	3	9
+	1,	7	4	5

4)

	6,	7	9	8
+	6,	4	3	4

5)

	8	4,	0	1	5
+		6,	2	4	3

6)

	2	6,	5	7	8
+	7	3,	5	9	4

7)

	5	6,	4	8	9
+	3	9,	2	6	3

8)

	9	6,	5	4	7
+	8	7,	9	6	0

9)

	4	8	3,	7	9	5
+	6	1	3,	7	0	8

10)

	7	9	6,	5	4	5
+	5	3	6,	8	2	6

11)

	2	5	4,	0	6	9
+	2	6	3,	0	7	8

12)

	9	9	4,	3	6	8
+			8,	9	2	7

Name: Date: Score: Unstoppable Owl

27) Adding Large Numbers

1)

	5,	3	7	1
+	3,	7	8	3

2)

	6,	5	7	3
+	9,	2	4	5

3)

	8,	0	6	5
+	8,	9	2	5

4)

	7,	8	4	6
+	4,	5	9	6

5)

	7	4,	2	3	0
+	9	6,	5	0	0

6)

	3	6,	8	5	4
+	3	7,	3	9	4

7)

	5	1,	9	5	8
+	6	2,	9	7	4

8)

		7,	4	5	7
+	5	6,	9	5	8

9)

	1	7	5,	3	8	5
+	2	6	0,	4	6	9

10)

	9	7	9,	4	8	3
+	4	8	5,	7	4	6

11)

	6	5	4,	9	8	5
+	8	7	4,	3	1	6

12)

	3	9	5,	8	7	1
+	4	7	6,	4	6	9

Name: Date: Score: Unstoppable Owl

28) Adding Large Numbers

1)

	5,	7	1	4
+	4,	6	3	5

2)

	4,	6	5	3
+	7,	8	9	7

3)

	4,	6	4	8
+	2,	6	9	7

4)

	9,	5	7	5
+	8,	4	7	3

5)

	7	9,	4	3	8
+			6	2	7

6)

	6	3,	4	7	8
+	5	9,	8	5	2

7)

	6	1,	9	7	4
+	5	3,	8	7	8

8)

	2	8,	9	4	0
+	5	3,	7	1	0

9)

	4	7	8,	4	1	5
+	1	3	0,	6	1	6

10)

	3	8	4,	9	6	4
+	9	8	6,	0	5	2

11)

	9	9	9,	9	9	9
+	9	8	7,	6	5	5

12)

	5	6	2,	7	8	6
+	5	8	2,	1	9	7

Name: Date: Score:

Unstoppable Owl

29) Adding Large Numbers

1)

	5,	1	7	5
+	4,	3	8	4

2)

	7,	9	5	6
+	4,	9	4	5

3)

	3,	8	7	3
+	2,	4	9	3

4)

	8,	4	6	5
+	9,	5	7	5

5)

	8	9,	6	5	6
+	2	4,	7	2	9

6)

	5	3,	7	5	9
+	4	4,	8	6	4

7)

	1	6,	8	4	0
+	1	7,	9	0	9

8)

	3	5,	8	7	9
+	7	3,	2	6	2

9)

	4	1	7,	4	3	9
+		8	8,	7	2	4

10)

	7	3	8,	5	9	6
+	7	9	1,	5	3	7

11)

	5	7	6,	0	0	5
+	4	3	8,	0	6	3

12)

	6	8	5,	2	7	8
+	1	2	6,	3	5	9

Name: Date: Score: Unstoppable Owl

30) Adding Large Numbers

1) **2,847 + 6,245 =**

2) **5,187 + 2,649 =**

3) **3,572 + 3,564 =**

4) **6,374 + 5,658 =**

5) **57,482 + 84,957 =**

6) **27,483 + 38,765 =**

7) **94,738 + 86,391 =**

8) **65,822 + 58,308 =**

9) **407,629 + 379,455 =**

10) **754,007 + 587,652 =**

Name: Date: Score:

Unstoppable Owl

31) Adding Large Numbers

1) **765 + 5,492 =**

2) **6,827 + 1,754 =**

3) **8,085 + 4,476 =**

4) **9,477 + 819 =**

5) **57,286 + 68,495 =**

6) **28,952 + 45,376 =**

7) **74,834 + 6,519 =**

8) **53,700 + 58,639 =**

9) **735,839 + 542,357 =**

10) **281,453 + 573,641 =**

Name: Date: Score:

Unstoppable Owl

32) Adding Large Numbers

1) **5,728 + 3,447 =**

2) **1,756 + 1,597 =**

3) **4,398 + 2,725 =**

4) **75 + 3,689 =**

5) **95,700 + 63,741 =**

6) **8,478 + 53,392 =**

7) **18,624 + 14,739 =**

8) **47,285 + 7,962 =**

9) **195,804 + 376,259 =**

10) **65,826 + 825,791 =**

Name: Date: Score: Unstopplable Owl

33) Adding Large Numbers

1) **6,287 + 4,193 =**

2) **573 + 2,649 =**

3) **5,748 + 2,354 =**

4) **7,412 + 9,563 =**

5) **46,539 + 17,265 =**

6) **90,672 + 26,854 =**

7) **56,270 + 968 =**

8) **43,847 + 7,625 =**

9) **541,677 + 586,364 =**

10) **472,849 + 493,423 =**

Name: Date: Score: *Unstoppable Owl*

34) Adding Large Numbers

1) **6,276 + 5,389 =**

2) **4,918 + 7,864 =**

3) **5,094 + 1,257 =**

4) **3,682 + 3,789 =**

5) **4,287 + 26,745 =**

6) **27,732 + 14,881 =**

7) **70,650 + 43,500 =**

8) **68,471 + 49,380 =**

9) **279,568 + 250,715 =**

10) **376,558 + 172,193 =**

Name: Date: Score: *Unstoppable Owl*

35) Double-Digit Subtraction (Without Regrouping)

Subtracting 2-digit numbers is similar to adding 2-digit numbers. Focus on one column at a time and subtract the digits in that column.

The problems in this section will not require regrouping. That skill will be learned later.

Step 1: Subtract the digits in the ones column. Write the answer directly below.
Step 2: Subtract the digits in the tens column and write that answer directly below.

Example:

	5	7
–	2	3

57 – 23 =

Step 1:

	5	7
–	2	3
		4

7 – 3 = **4**

Step 2:

	5	7
–	2	3
	3	4

5 – 2 = **3**

Answer:

	5	7
–	2	3
	3	4

57 – 23 = **34**

Directions: *Solve each subtraction problem.*
Remember to subtract the digits in each column, one column at a time.

1)

	6	5
–	1	4

2)

	7	2
–	4	2

3)

	3	5
–	3	1

4)

	9	7
–	1	3

5)

	4	1
–	2	0

6)

	8	3
–	7	2

7)

	6	8
–	2	5

8)

	2	9
–	1	1

9)

	7	8
–	6	2

10)

	5	7
–	5	6

11)

	7	8
–	6	0

12)

	3	4
–	1	4

Name: Date: Score: *Unstoppable Owl*

36) Double-Digit Subtraction (Without Regrouping)

Problem	Top	Subtract
1)	94	− 13
2)	68	− 36
3)	46	− 22
4)	57	− 10
5)	35	− 31
6)	79	− 24
7)	68	− 52
8)	53	− 21
9)	89	− 25
10)	45	− 14
11)	79	− 72
12)	65	− 43
13)	88	− 17
14)	56	− 23
15)	92	− 61
16)	60	− 20
17)	97	− 21
18)	75	− 52
19)	48	− 14
20)	53	− 23
21)	86	− 35
22)	65	− 15
23)	28	− 21
24)	74	− 62

Name: Date: Score:

Unstoppable Owl

37) Double-Digit Subtraction (Without Regrouping)

1)

	5	7
−	2	2

2)

	6	2
−	1	2

3)

	8	8
−	3	6

4)

	5	3
−	1	2

5)

	7	6
−	3	2

6)

	4	0
−	1	0

7)

	2	5
−	1	4

8)

	4	9
−	2	1

9)

	8	5
−	2	5

10)

	9	8
−	1	4

11)

	4	3
−	3	1

12)

	6	7
−	2	5

13)

	9	4
−	5	3

14)

	7	7
−	1	4

15)

	4	8
−	4	3

16)

	5	6
−	3	1

17)

	6	4
−	3	1

18)

	5	9
−	4	6

19)

	3	3
−	2	3

20)

	7	8
−	5	2

21)

	4	7
−	2	4

22)

	6	1
−	4	0

23)

	5	9
−	3	8

24)

	9	9
−	3	1

Name: Date: Score: *Unstoppable Owl*

38) Double-Digit Subtraction (Without Regrouping)

1) $49 - 32$

2) $68 - 43$

3) $57 - 12$

4) $93 - 71$

5) $65 - 51$

6) $74 - 12$

7) $89 - 49$

8) $72 - 30$

9) $83 - 31$

10) $59 - 24$

11) $33 - 21$

12) $86 - 15$

13) $75 - 33$

14) $99 - 52$

15) $47 - 12$

16) $85 - 71$

17) $26 - 14$

18) $74 - 24$

19) $55 - 14$

20) $28 - 13$

21) $73 - 43$

22) $95 - 21$

23) $87 - 44$

24) $64 - 23$

Unstoppable Owl

Name: Date: Score:

39) Double-Digit Subtraction (Without Regrouping)

1) $58 - 25$

2) $97 - 14$

3) $64 - 32$

4) $38 - 21$

5) $49 - 36$

6) $58 - 40$

7) $27 - 12$

8) $65 - 41$

9) $73 - 21$

10) $88 - 56$

11) $67 - 23$

12) $59 - 33$

13) $94 - 14$

14) $68 - 52$

15) $45 - 24$

16) $74 - 30$

17) $55 - 32$

18) $93 - 21$

19) $68 - 42$

20) $49 - 44$

21) $80 - 30$

22) $52 - 41$

23) $47 - 12$

24) $75 - 53$

Name: Date: Score: *Unstoppable Owl*

40) **Double-Digit Subtraction** (With Regrouping)

When regrouping is done while *adding*, an excess amount is transferred from one column to the next. When regrouping during *subtraction*, the opposite occurs. If the top digit in the column has a value that is less than the value of the bottom digit, regrouping is needed. Part of the value from a column to the left can be transferred to the right and added to the top digit.

In the example below. The 2 in the ones column is less than the 7. To borrow from the tens column, 1 of the tens will be moved from the tens column to the ones column. If there were 5 tens, and 1 ten is taken away, there will only be 4 tens left in the tens column. The ten being regrouped is then added to the ones column. 10 + 2 = 12, so the value in the ones column becomes 12.

12 is greater than 7, so subtraction in the ones column can now take place. 12 – 7 = 5.

Subtraction in the tens column is next. 4 – 1 = 3. Therefore, 52 – 17 = 35.

Step 1: Subtract the digits in the ones column.
(*If necessary*, regroup from the tens column to the ones column first.)
Step 2: Subtract the digits in the tens column.

Step 1: Regroup

	4	
	~~5~~	12
–	1	7

2 is less than 7.
Regroup first.

Step 1: Subtract

	4	
	~~5~~	12
–	1	7
		5

12 – 7 = **5**

Step 2: Subtract

	4	
	~~5~~	12
–	1	7
	3	5

4 – 1 = **3**

Answer:

	4	
	~~5~~	12
–	1	7
	3	5

52 – 17 = **35**

Directions: *Solve each subtraction problem.* ***Many problems will require regrouping, but some may not.***

1)

	8	4
–	3	9

2)

	5	6
–	1	7

3)

	4	0
–	1	6

4)

	9	5
–	2	7

5)

	5	7
–	4	3

6)

	6	8
–	3	9

7)

	4	2
–	1	3

8)

	7	3
–	2	8

Name: Date: Score: Unstoppable Owl

41) Double-Digit Subtraction

1) 75 − 57 =

2) 48 − 29 =

3) 56 − 48 =

4) 80 − 22 =

5) 51 − 34 =

6) 95 − 17 =

7) 32 − 7 =

8) 63 − 59 =

9) 74 − 26 =

10) 57 − 38 =

11) 64 − 36 =

12) 85 − 48 =

13) 44 − 29 =

14) 72 − 35 =

15) 61 − 33 =

16) 93 − 47 =

17) 85 − 36 =

18) 27 − 8 =

19) 47 − 19 =

20) 68 − 26 =

21) 76 − 49 =

22) 32 − 16 =

23) 53 − 28 =

24) 61 − 13 =

Name: Date: Score: Unstoppable Owl

42) Double-Digit Subtraction

1)

	5	4
−	2	7

2)

	8	2
−	3	9

3)

	7	1
−	4	6

4)

	4	3
−	1	5

5)

	9	0
−	6	7

6)

	7	5
−	2	8

7)

	4	8
−	3	9

8)

	8	4
−	5	8

9)

	6	2
−	3	9

10)

	4	0
−	1	6

11)

	8	3
−	4	7

12)

	5	1
−	3	8

13)

	6	4
−	2	5

14)

	7	5
−	2	8

15)

	3	2
−	1	4

16)

	7	3
−	1	9

17)

	5	5
−	1	7

18)

	8	6
−	5	8

19)

	6	1
−	1	7

20)

	4	6
−	2	7

21)

	7	7
−	5	9

22)

	5	0
−	2	1

23)

	8	5
−	4	9

24)

	9	4
−	3	7

Name: Date: Score: Unstoppable Owl

43) Double-Digit Subtraction

1) 74 − 47 =

2) 58 − 16 =

3) 42 − 7 =

4) 61 − 34 =

5) 37 − 19 =

6) 65 − 28 =

7) 54 − 39 =

8) 82 − 64 =

9) 71 − 36 =

10) 86 − 29 =

11) 32 − 18 =

12) 43 − 25 =

13) 53 − 6 =

14) 55 − 28 =

15) 86 − 47 =

16) 50 − 18 =

17) 72 − 35 =

18) 13 − 5 =

19) 61 − 47 =

20) 94 − 76 =

21) 60 − 17 =

22) 53 − 39 =

23) 85 − 27 =

24) 72 − 53 =

Name: Date: Score: Unstoppable Owl

44) Double-Digit Subtraction

1) 47 − 19

2) 82 − 75

3) 65 − 23

4) 57 − 28

5) 73 − 55

6) 31 − 14

7) 94 − 28

8) 45 − 27

9) 82 − 26

10) 50 − 33

11) 76 − 18

12) 63 − 39

13) 20 − 16

14) 64 − 51

15) 80 − 59

16) 15 − 8

17) 93 − 16

18) 75 − 49

19) 61 − 17

20) 32 − 14

21) 70 − 32

22) 54 − 28

23) 69 − 31

24) 46 − 39

Name: Date: Score: Unstoppable Owl

45) Double-Digit Subtraction

You've had a chance to practice subtracting double-digit numbers using guidelines. Now it's time to write the problems yourself. Remember to align the columns and place values.

(Refer to page 11 if you would like to review these skills.)

Directions: *Solve each subtraction problem.*
Many problems will require regrouping, but regrouping may not be necessary.

1) **56 – 28 =**

2) **74 – 39 =**

3) **62 – 33 =**

4) **54 – 35 =**

5) **94 – 57 =**

6) **44 – 18 =**

7) **71 – 53 =**

8) **63 – 27 =**

9) **90 – 53 =**

10) **85 – 36 =**

11) **47 – 29 =**

12) **42 – 16 =**

13) **83 – 63 =**

14) **71 – 26 =**

15) **50 – 22 =**

16) **41 – 12 =**

Name: Date: Score: Unstoppable Owl

46) Double-Digit Subtraction

1) 95 – 47 =
2) 24 – 15 =
3) 63 – 39 =
4) 56 – 27 =

5) 76 – 23 =
6) 72 – 8 =
7) 91 – 34 =
8) 55 – 28 =

9) 64 – 26 =
10) 81 – 65 =
11) 53 – 26 =
12) 42 – 25 =

13) 47 – 9 =
14) 36 – 17 =
15) 84 – 49 =
16) 65 – 15 =

17) 92 – 27 =
18) 61 – 43 =
19) 73 – 25 =
20) 54 – 27 =

Name: Date: Score: Unstoppable Owl

47) Double-Digit Subtraction

1) **81 – 47 =** 2) **64 – 35 =** 3) **73 – 28 =** 4) **92 – 45 =**

5) **60 – 31 =** 6) **51 – 24 =** 7) **94 – 77 =** 8) **55 – 6 =**

9) **46 – 25 =** 10) **84 – 38 =** 11) **32 – 14 =** 12) **53 – 26 =**

13) **25 – 17 =** 14) **35 – 19 =** 15) **93 – 29 =** 16) **41 – 23 =**

17) **56 – 38 =** 18) **62 – 34 =** 19) **87 – 18 =** 20) **65 – 27 =**

Name: Date: Score: Unstoppable Owl

48) Double-Digit Subtraction

1) **58 – 26 =** 2) **40 – 27 =** 3) **83 – 35 =** 4) **44 – 16 =**

5) **73 – 48 =** 6) **91 – 59 =** 7) **84 – 8 =** 8) **62 – 13 =**

9) **75 – 46 =** 10) **93 – 37 =** 11) **52 – 34 =** 12) **70 – 43 =**

13) **63 – 39 =** 14) **86 – 58 =** 15) **65 – 28 =** 16) **41 – 17 =**

17) **50 – 46 =** 18) **67 – 49 =** 19) **58 – 56 =** 20) **72 – 47 =**

Name: Date: Score: Unstoppable Owl

49) Double-Digit Subtraction

1) **52 – 23 =**	2) **74 – 38 =**	3) **81 – 35 =**	4) **95 – 17 =**
5) **43 – 35 =**	6) **65 – 29 =**	7) **72 – 7 =**	8) **46 – 28 =**
9) **87 – 48 =**	10) **91 – 59 =**	11) **63 – 28 =**	12) **52 – 24 =**
13) **60 – 26 =**	14) **83 – 4 =**	15) **48 – 19 =**	16) **74 – 25 =**
17) **74 – 34 =**	18) **26 – 19 =**	19) **54 – 27 =**	20) **92 – 45 =**

Name: Date: Score: *Unstoppable Owl*

50) Triple-Digit Subtraction

Subtracting 3-digit numbers is similar to subtracting 2-digit numbers. There's just an extra column to subtract—the hundreds column. Sometimes, you may need to regroup from the tens column to the ones column, and you may need to regroup from the hundreds column to the tens column.

Step 1: Subtract the ones column. (Regroup from the tens to ones column *if necessary*.)
Step 2: Subtract the tens column. (Regroup from the hundreds to tens column *if necessary*.)
Step 3: Subtract the digits in the hundreds column.

Example: 750 – 285 =

Step 1:

		4	
	7	~~5~~	10
–	2	8	5
			5

Regroup
10 – 5 = **5**

Step 2:

	6	14	
	~~7~~	~~5~~	0
–	2	8	5
		6	5

Regroup
14 – 8 = **6**

Step 3:

	6	14	
	~~7~~	~~5~~	0
–	2	8	5
	4	6	5

6 – 2 = **4**
Answer: 750 – 285 = **465**

Directions: *Solve each subtraction problem.*
Many problems will require regrouping in one or more columns, but regrouping may not be necessary.

1)

	4	6	5
–	3	8	1

2)

	8	5	0
–	4	1	6

3)

	2	5	3
–	1	9	7

4)

	9	4	5
–	3	4	8

5)

	5	2	6
–	1	7	7

6)

	7	6	1
–	4	3	2

7)

	6	5	3
–		7	9

8)

	3	4	2
–	1	6	8

9)

	5	7	4
–	3	3	5

Name: Date: Score: *Unstoppable Owl*

51) Triple-Digit Subtraction

1)

	7	2	4
–	1	5	9

2)

	9	3	4
–	5	5	7

3)

	6	1	3
–	4	3	4

4)

	8	4	5
–	3	4	1

5)

	6	5	3
–	1	5	8

6)

	9	4	2
–	8	7	0

7)

	6	3	2
–	2	5	8

8)

	7	1	0
–	1	2	7

9)

	5	6	3
–	4	9	2

10)

	3	5	7
–		6	8

11)

	4	7	3
–	2	6	9

12)

	5	1	1
–	2	4	7

13)

	2	6	8
–	1	7	3

14)

	7	3	5
–	5	4	8

15)

	6	5	2
–	3	7	7

16)

	8	2	4
–	2	3	6

17)

	9	1	6
–	5	1	7

18)

	4	7	0
–	3	5	6

Name: Date: Score: *Unstoppable Owl*

52) Triple-Digit Subtraction (Regrouping with a 0)

Sometimes, regrouping is necessary, but the digit in the next column is a 0. When this happens, keep moving to columns further left to regroup from. In the example below, the tens column has a 0. Before you can regroup from the tens column to the ones column, you first need to regroup from the hundreds column to the tens column, making the value of the tens column change from 0 to 10. At that point, regrouping from the tens column to the ones column can take place.

Step 1: There is a 0 in the ten's column, so regrouping from the hundred's column to the ten's column must be done first.
Step 2: Now regrouping from the ten's column to the one's column can take place.
Step 3: The regrouping is done. Subtraction can take place in each column.

Example: 604 – 278 =

Step 1:

	5		
	~~6~~	10	4
–	2	7	8

Regroup
(Hundreds to tens column.)

Step 2:

	5	9	
	~~6~~	~~10~~	14
–	2	7	8

Regroup
(Tens to ones column.)

Step 3:

	5	9	
	~~6~~	~~10~~	14
–	2	7	8
	3	2	6

Subtract in each column.
Answer: 604 – 278 = **326**

Directions: *Solve each subtraction problem.*
Many problems will require regrouping in one or more columns, but regrouping may not be necessary.

1)

	4	0	6
–	1	2	7

2)

	5	0	2
–	3	4	4

3)

	9	0	0
–	4	5	3

4)

	6	0	7
–	5	7	9

5)

	8	0	1
–	2	9	5

6)

	3	0	8
–	2	1	9

Name: Date: Score: Unstoppable Owl

53) Triple-Digit Subtraction

1)

	5	4	1
–	1	7	3

2)

	8	0	4
–	4	3	6

3)

	6	5	1
–	2	9	4

4)

	5	0	0
–	2	1	6

5)

	4	2	7
–	1	5	9

6)

	7	0	5
–	4	2	6

7)

	8	5	1
–	5	4	7

8)

	2	6	4
–	2	2	8

9)

	3	5	2
–	1	9	4

10)

	5	1	3
–	3	6	7

11)

	4	0	0
–	1	8	2

12)

	7	4	6
–	3	5	9

13)

	6	4	7
–	2	4	3

14)

	8	4	5
–	5	6	7

15)

	5	0	1
–	1	2	2

16)

	7	0	4
–	3	3	8

17)

	5	6	3
–	3	7	2

18)

	9	3	2
–	2	6	8

Name: Date: Score: Unstoppable Owl

54) Triple-Digit Subtraction

1)

	6	2	1
−	1	2	5

2)

	7	8	5
−		6	2

3)

	3	0	3
−	1	3	5

4)

	7	3	6
−	5	4	2

5)

	6	0	0
−	2	3	9

6)

	2	6	1
−		9	2

7)

	5	0	4
−	2	7	7

8)

	9	4	3
−	4	6	8

9)

	3	7	6
−	1	9	4

10)

	4	1	8
−	1	5	9

11)

	3	8	2
−	3	4	6

12)

	8	0	0
−	2	1	6

13)

	6	9	5
−	2	9	6

14)

	7	0	2
−	1	0	8

15)

	6	3	3
−	5	5	9

16)

	9	0	0
−	5	7	3

17)

	2	4	5
−		7	8

18)

	4	3	7
−	1	3	8

Name: Date: Score: Unstoppable Owl

55) Triple-Digit Addition (Without Guidelines)

It's time to practice writing triple-digit subtraction problems yourself. Remember to align the columns and place values. (Refer to page 11 if you would like to review these skills.)

1) **267 – 184 =**

2) **525 – 246 =**

3) **903 – 465 =**

4) **137 – 75 =**

5) **472 – 294 =**

6) **700 – 261 =**

7) **504 – 326 =**

8) **413 – 265 =**

9) **780 – 367 =**

10) **473 – 183 =**

11) **524 – 335 =**

12) **721 – 79 =**

Name: Date: Score: Unstopppable Owl

56) Triple-Digit Subtraction

1) **642 – 378 =**

2) **361 – 288 =**

3) **200 – 37 =**

4) **703 – 346 =**

5) **945 – 261 =**

6) **574 – 194 =**

7) **356 – 178 =**

8) **581 – 429 =**

9) **306 – 253 =**

10) **400 – 362 =**

11) **917 – 563 =**

12) **451 – 178 =**

13) **802 – 425 =**

14) **492 – 467 =**

15) **100 – 74 =**

Name: Date: Score: *Unstoppable Owl*

57) Triple-Digit Subtraction

1) **384 – 155 =**

2) **803 – 416 =**

3) **731 – 256 =**

4) **500 – 176 =**

5) **635 – 149 =**

6) **506 – 207 =**

7) **771 – 278 =**

8) **605 – 208 =**

9) **300 – 181 =**

10) **450 – 237 =**

11) **792 – 593 =**

12) **612 – 345 =**

13) **900 – 334 =**

14) **627 – 185 =**

15) **681 – 292 =**

Name: Date: Score: Unstoppable Owl

58) Triple-Digit Subtraction

1) **300 – 216 =**

2) **738 – 456 =**

3) **622 – 364 =**

4) **903 – 654 =**

5) **672 – 179 =**

6) **366 – 289 =**

7) **720 – 650 =**

8) **400 – 91 =**

9) **238 – 160 =**

10) **492 – 319 =**

11) **441 – 253 =**

12) **623 – 276 =**

13) **834 – 168 =**

14) **557 – 129 =**

15) **406 – 278 =**

Name: Date: Score: Unstoppable Owl

59) Triple-Digit Subtraction

1) **472 – 189 =**

2) **500 – 261 =**

3) **825 – 176 =**

4) **541 – 377 =**

5) **992 – 256 =**

6) **726 – 395 =**

7) **700 – 143 =**

8) **680 – 415 =**

9) **536 – 519 =**

10) **637 – 84 =**

11) **406 – 178 =**

12) **394 – 165 =**

13) **302 – 182 =**

14) **472 – 89 =**

15) **731 – 348 =**

Name: Date: Score: Unstoppable Owl

60) Subtracting Large Numbers

It's time to focus on subtracting large numbers. The process of subtracting each column and borrowing if needed still applies. If borrowing is needed and a 0 is in the column, remember that you will need to first borrow from a column further left (as was done in example 2).

Example 1: 8,156 – 5,429 =

	7		4	
	~~8~~,	11	~~5~~	16
–	5,	4	2	9
	2,	7	2	7

8,156 – 5,429 = **2,727**

Example 2: 480,047 – 376,250 =

		7	9	9		
	4	~~8~~	~~10~~,	~~10~~	14	7
–	3	7	6,	2	5	0
	1	0	3,	7	9	7

480,047 – 376,250 = **103,797**

Directions: *Solve each addition problem.*

Many problems will require regrouping in one or more columns, but regrouping may not be necessary.

1)

	3,	5	8	0
–	1,	6	4	2

2)

	8,	9	0	3
–	8,	4	5	8

3)

	4	0,	3	1	4
–		2,	7	0	6

4)

	5	4,	2	8	3
–	1	7,	5	5	4

5)

	9	1	6,	7	1	4
–	5	2	3,	9	5	7

6)

	6	3	2,	0	3	5
–	5	5	1,	9	7	3

Name: Date: Score: *Unstoppable Owl*

61) Subtracting Large Numbers

1)

	5,	0	0	0
−	2,	3	1	6

2)

	7,	2	5	3
−	6,	1	9	4

3)

	3,	6	3	3
−		6	8	1

4)

	4,	0	0	2
−	1,	8	3	5

5)

	2	0,	8	7	5
−	2	0,	4	7	6

6)

	6	1,	3	7	8
−		9,	5	4	8

7)

	3	2,	6	0	3
−	1	7,	9	3	8

8)

	4	0,	7	3	8
−	2	4,	5	1	9

9)

	5	1	2,	9	4	0
−	3	1	4,	7	6	0

10)

	4	7	1,	0	6	2
−	1	8	4,	2	9	8

11)

	7	2	6,	1	6	8
−	3	5	3,	2	5	7

12)

	9	0	4,	7	5	8
−	2	5	5,	9	2	1

Name: Date: Score: Unstoppable Owl

62) Subtracting Large Numbers

1)

	5,	7	2	4
−	5,	2	8	6

2)

	7,	1	3	6
−	1,	0	4	9

3)

	3,	2	5	7
−	2,	6	3	9

4)

	7,	0	0	0
−		5	1	8

5)

	3	0,	0	5	8
−		2,	1	9	2

6)

	5	4,	3	2	0
−	1	9,	7	5	1

7)

	4	7,	5	1	3
−	2	1,	6	8	5

8)

	6	2,	7	5	1
−	3	2,	9	3	4

9)

	1	0	0,	0	0	0
−		3	9,	2	6	5

10)

	4	1	3,	6	9	2
−	1	3	9,	7	9	3

11)

	5	6	7,	3	0	0
−	2	4	9,	5	1	4

12)

	6	0	0,	2	5	0
−	4	3	6,	5	2	7

Name: Date: Score: Unstoppable Owl

63) Subtracting Large Numbers

1)

	6,	3	0	0
−	1,	5	2	8

2)

	3,	7	1	5
−	2,	4	8	6

3)

	5,	6	0	3
−	1,	9	3	5

4)

	6,	0	7	5
−		5	4	1

5)

	8	3,	0	0	6
−	7	5,	4	1	4

6)

	2	7,	3	8	1
−	1	9,	6	5	2

7)

	6	2,	6	0	0
−		3,	2	4	8

8)

	5	1,	4	8	7
−	2	3,	5	7	2

9)

	2	0	6,	5	1	0
−	1	3	8,	0	6	4

10)

	3	7	2,	5	4	3
−	1	6	9,	7	5	3

11)

	9	5	3,	7	2	8
−	4	6	9,	1	0	9

12)

	6	0	0,	2	8	5
−	5	3	4,	6	5	2

Name: Date: Score: Unstoppable Owl

64) Subtracting Large Numbers

1)

	7,	2	6	1
–	3,	4	5	5

2)

	4,	5	7	8
–		8	9	4

3)

	8,	0	0	0
–	2,	1	3	7

4)

	6,	1	4	3
–	1,	4	8	6

5)

	3	6,	3	0	4
–	1	6,	5	2	6

6)

	2	0,	4	8	5
–			9	3	6

7)

	8	9,	6	2	4
–	4	7,	3	1	4

8)

	5	0,	0	0	0
–	3	4,	0	8	2

9)

	4	1	7,	5	5	9
–	1	3	6,	5	8	2

10)

	9	1	2,	6	4	2
–	2	7	3,	9	1	3

11)

	4	3	5,	7	1	1
–		4	7,	4	9	0

12)

	6	2	8,	5	2	8
–	5	6	1,	7	5	3

Name: Date: Score: Unstoppable Owl

65) Subtracting Large Numbers

Practice writing large subtraction problems yourself. Carefully align the columns and place values before subtracting. (Refer to page 11 if you would like to review these skills.)

1) **7,135 – 2,844 =**

2) **9,000 – 3,041 =**

3) **5,286 – 719 =**

4) **4,052 – 1,256 =**

5) **50,627 – 23,154 =**

6) **26,470 – 7,947 =**

7) **700,000 – 361,429 =**

8) **325,715 – 146,348 =**

Name: Date: Score: Unstoppable Owl

66) Subtracting Large Numbers

1) **2,600 – 1,372 =**

2) **8,352 – 2,794 =**

3) **6,475 – 4,855 =**

4) **5,000 – 86 =**

5) **94,360 – 37,598 =**

6) **63,725 – 54,167 =**

7) **40,059 – 18,163 =**

8) **75,000 – 11,430 =**

9) **682,538 – 264,773 =**

10) **240,318 – 87,572 =**

Name: Date: Score: Unstoppable Owl

67) Subtracting Large Numbers

1) **3,510 – 3,265 =**

2) **6,379 – 1,546 =**

3) **8,100 – 4,935 =**

4) **7,153 – 6,379 =**

5) **45,623 – 15,875 =**

6) **70,473 – 41,560 =**

7) **82,530 – 56,174 =**

8) **10,000 – 6,389 =**

9) **420,165 – 125,853 =**

10) **725,000 – 418,295 =**

Name: Date: Score: Unstoppable Owl

68) Subtracting Large Numbers

1) **8,347 – 3,518 =**

2) **4,178 – 1,999 =**

3) **2,307 – 1,468 =**

4) **6,003 – 4,615 =**

5) **39,284 – 37,609 =**

6) **71,382 – 42,185 =**

7) **76,300 – 15,787 =**

8) **92,146 – 6,570 =**

9) **415,863 – 240,785 =**

10) **600,000 – 148,925 =**

Name: Date: Score:

69) Subtracting Large Numbers

1) **7,046 – 3,355 =**

2) **5,178 – 1,259 =**

3) **9,050 – 2,415 =**

4) **6,421 – 5,683 =**

5) **51,638 – 24,975 =**

6) **18,263 – 11,436 =**

7) **72,952 – 35,724 =**

8) **31,000 – 14,978 =**

9) **623,541 – 268,970 =**

10) **512,803 – 367,495 =**

Name: Date: Score: Unstoppable Owl

70) Mixed Review

1) 56 + 75 =

2) 94 + 83 =

3) 78 + 6 =

4) 16 + 98 =

5) 6,409 + 7,592 =

6) 37,832 + 45,497 =

7) 61,472 + 55,348 =

8) 75 – 38 =

9) 24 – 9 =

10) 90 – 42 =

11) 31 – 14 =

12) 5,152 – 1,657 =

13) 40,000 – 21,395 =

14) 83,637 – 54,852 =

Name: Date: Score: Unstoppable Owl

71) Mixed Review

1) **70 + 42 =**

2) **65 + 85 =**

3) **48 + 94 =**

4) **36 + 57 =**

5) **2,187 + 9,446 =**

6) **37,539 + 85,642 =**

7) **99,856 + 82,726 =**

8) **57 – 26 =**

9) **83 – 49 =**

10) **46 – 27 =**

11) **70 – 35 =**

12) **5,714 – 2,395 =**

13) **60,200 – 21,365 =**

14) **84,946 – 37,284 =**

Name: Date: Score:

Unstoppable Owl

72) Mixed Review

1) **86 + 38 =**

2) **29 + 73 =**

3) **97 + 64 =**

4) **7 + 35 =**

5) **2,897 + 2,736 =**

6) **9,276 + 93,843 =**

7) **27,319 + 15,473 =**

8) **30 – 21 =**

9) **82 – 17 =**

10) **75 – 36 =**

11) **64 – 39 =**

12) **7,004 – 215 =**

13) **36,021 – 19,462 =**

14) **94,432 – 34,857 =**

Name: Date: Score: Unstoppable Owl

73) Mixed Review

1) **77 + 46 =**

2) **84 + 95 =**

3) **57 + 65 =**

4) **29 + 53 =**

5) **8,573 + 4,924 =**

6) **58,750 + 65,589 =**

7) **29,671 + 584 =**

8) **50 – 18 =**

9) **62 – 39 =**

10) **74 – 54 =**

11) **81 – 43 =**

12) **6,125 – 3,754 =**

13) **49,520 – 16,783 =**

14) **10,156 – 1,462 =**

Name: Date: Score: Unstoppable Owl

74) Mixed Review

1) **47 + 47 =**

2) **61 + 75 =**

3) **38 + 92 =**

4) **59 + 63 =**

5) **4,724 + 7,085 =**

6) **37,591 + 88,925 =**

7) **26,781 + 26,453 =**

8) **57 – 38 =**

9) **74 – 45 =**

10) **93 – 26 =**

11) **86 – 25 =**

12) **5,200 – 3,941 =**

13) **47,603 – 28,174 =**

14) **80,000 – 45,027 =**

Name: Date: Score:

Unstoppable Owl

75) Mixed Review

1) **18 + 27 =**

2) **63 + 29 =**

3) **37 + 94 =**

4) **55 + 55 =**

5) **2,627 + 1,815 =**

6) **46,943 + 65,927 =**

7) **25,875 + 25,659 =**

8) **41 – 16 =**

9) **53 – 27 =**

10) **76 – 58 =**

11) **32 – 5 =**

12) **4,036 – 1,352 =**

13) **31,758 – 14,279 =**

14) **70,521 – 16,837 =**

Name: Date: Score: Unstoppable Owl

76) Mixed Review

1) **77 + 77 =**

2) **65 + 89 =**

3) **16 + 95 =**

4) **87 + 46 =**

5) **8,370 + 5,657 =**

6) **47,573 + 39,166 =**

7) **98,375 + 76,945 =**

8) **52 – 15 =**

9) **73 – 26 =**

10) **80 – 31 =**

11) **44 – 28 =**

12) **6,374 – 5,571 =**

13) **51,700 – 8,926 =**

14) **12,738 – 4,576 =**

Name: Date: Score: Unstoppable Owl

77) Mixed Review

1) 64 + 78 =

2) 97 + 99 =

3) 27 + 34 =

4) 56 + 86 =

5) 3,542 + 6,759 =

6) 18,532 + 23,794 =

7) 45,187 + 27,054 =

8) 53 – 16 =

9) 75 – 28 =

10) 92 – 75 =

11) 47 – 23 =

12) 8,471 – 2,492 =

13) 35,000 – 17,425 =

14) 68,275 – 59,428 =

Name: Date: Score: Unstoppable Owl

78) Mixed Review

1) **27 + 84 =**

2) **94 + 53 =**

3) **35 + 35 =**

4) **64 + 21 =**

5) **3,639 + 1,465 =**

6) **37,491 + 68,726 =**

7) **22,755 + 24,684 =**

8) **83 – 67 =**

9) **60 – 36 =**

10) **46 – 28 =**

11) **54 – 15 =**

12) **7,140 – 6,463 =**

13) **14,752 – 5,094 =**

14) **50,700 – 13,472 =**

Name: Date: Score: Unstoppable Owl

79) Mixed Review

1) **23 + 79 =**

2) **67 + 45 =**

3) **72 + 54 =**

4) **39 + 29 =**

5) **6,683 + 5,475 =**

6) **51,864 + 60,000 =**

7) **74,346 + 85,984 =**

8) **54 – 38 =**

9) **27 – 9 =**

10) **63 – 24 =**

11) **91 – 29 =**

12) **3,000 – 1,215 =**

13) **75,632 – 16,852 =**

14) **18,532 – 17,683 =**

Name: Date: Score: Unstoppable Owl

80) Mixed Review

1) **65 + 97 =**

2) **38 + 23 =**

3) **77 + 55 =**

4) **46 + 85 =**

5) **9,300 + 4,500 =**

6) **47,624 + 84,957 =**

7) **25,317 + 56,523 =**

8) **40 – 12 =**

9) **75 – 46 =**

10) **61 – 39 =**

11) **82 – 33 =**

12) **1,374 – 1,315 =**

13) **49,208 – 15,683 =**

14) **90,042 – 37,168 =**

Name: Date: Score: Unstoppable Owl

81) Mixed Review

1) **98 + 79 =**

2) **45 + 76 =**

3) **42 + 16 =**

4) **64 + 89 =**

5) **6,475 + 3,841 =**

6) **47,163 + 21,534 =**

7) **47,854 + 66,378 =**

8) **65 – 34 =**

9) **73 – 27 =**

10) **32 – 13 =**

11) **56 – 48 =**

12) **2,638 – 1,457 =**

13) **26,572 – 18,743 =**

14) **85,413 – 47,865 =**

Name: Date: Score: Unstoppable Owl

82) Mixed Review

1) **58 + 67 =**

2) **84 + 39 =**

3) **46 + 87 =**

4) **93 + 54 =**

5) **1,537 + 1,965 =**

6) **76,372 + 48,531 =**

7) **30,358 + 14,634 =**

8) **73 – 46 =**

9) **41 – 15 =**

10) **99 – 59 =**

11) **85 – 37 =**

12) **7,641 – 3,952 =**

13) **64,732 – 45,185 =**

14) **33,562 – 16,853 =**

Name: Date: Score: Unstoppable Owl

83) Mixed Review

1) **45 + 78 =**

2) **53 + 26 =**

3) **86 + 95 =**

4) **74 + 49 =**

5) **4,054 + 1,479 =**

6) **17,634 + 25,584 =**

7) **76,583 + 94,756 =**

8) **47 – 28 =**

9) **86 – 39 =**

10) **72 – 56 =**

11) **80 – 21 =**

12) **3,572 – 3,167 =**

13) **70,526 – 35,842 =**

14) **49,736 – 27,849 =**

Name: Date: Score:

Unstoppable Owl

84) Mixed Review

1) $67 + 25 =$

2) $73 + 26 =$

3) $56 + 17 =$

4) $92 + 54 =$

5) $6,571 + 8,735 =$

6) $56,824 + 76,953 =$

7) $15,572 + 38,974 =$

8) $37 - 19 =$

9) $45 - 32 =$

10) $64 - 35 =$

11) $91 - 13 =$

12) $9,000 - 274 =$

13) $62,153 - 25,674 =$

14) $36,428 - 32,152 =$

Name: Date: Score: Unstoppable Owl

85) Mixed Review

1) 76 + 44 =

2) 57 + 85 =

3) 39 + 68 =

4) 64 + 94 =

5) 7,368 + 4,572 =

6) 62,450 + 94,178 =

7) 37,261 + 8,794 =

8) 63 – 54 =

9) 51 – 37 =

10) 82 – 16 =

11) 45 – 8 =

12) 5,472 – 1,605 =

13) 75,000 – 26,831 =

14) 31,762 – 16,975 =

Name: Date: Score: Unstoppable Owl

86) 1-Step Word Problems

Knowing ***how*** to add and subtract are important skills, but even more important is knowing ***when*** to add or subtract.

The word problems in this section are designed to portray real-world situations. Pay close attention to the wording of these problems. They often contain **keywords** that may help to indicate whether addition or subtraction is needed to solve the problem.

Common Addition Keywords:		Common Subtraction Keywords:	
- Altogether	**- Plus**	**- Fewer**	**- How many more?**
- Total	**- More**	**- Less**	- How many **remain?**
- Combined	**- Increase**	**- Minus**	- How many are **left?**
- In all		**- Difference**	

1) **Kelly planted 58 seeds in one row and 45 seeds in another row. How many seeds were planted in both rows combined?**

2) **Kelly has 37 tomato plants and 60 bean plants. How many more bean plants does she have than tomato plants?**

3) **Kelly had 105 carrots. Then, rabbits ate 29 of the carrots. How many carrots are left?**

4) **Kelly picked 46 tomatoes in the morning and 34 tomatoes in the afternoon. How many tomatoes does that total?**

5) **Alex harvested 258 potatoes from his garden. Kelly harvested 173 potatoes from her garden. How many potatoes did they harvest altogether?**

6) **Alex harvested 258 potatoes, then gave 175 of them to his neighbors. How many potatoes does Alex have left?**

Name: Date: Score: Unstoppable Owl

87) 1-Step Word Problems

1) Ben owns a coffee shop. He sold 53 more cups of coffee today than yesterday. If he sold 793 cups of coffee yesterday, how many cups of coffee did he sell today?

2) This morning, he sold 314 large cups of coffee and 265 small cups of coffee. How many more large cups of coffee did he sell than small cups of coffee?

3) Ben's coffee shop used 126 bags of coffee beans last week and 138 bags of coffee beans this week. How many bags is that for both weeks combined?

4) Ben's coffee shop only has 37 bags of coffee beans left, so he had 285 more bags delivered. How many bags of coffee beans does the shop have now?

5) The coffee shop has 45 chairs for customers to sit on while they drink their coffee. Only 9 of the chairs are currently not being used by customers. How many chairs are being used by customers?

6) Ben does not think 45 chairs is enough. If he buys 12 more chairs for the coffee shop, how many chairs will there be altogether?

7) The coffee shop had 56 customers during the first hour it was open today, and 147 customers during the second hour it was open. How many customers was that during these two hours combined?

8) The coffee shop had 937 customers on Friday and 1,168 customers on Saturday. How many more customers were there on Saturday than Friday?

Name: Date: Score:

Unstoppable Owl

88) 1-Step Word Problems

1) **A farmer had 312 cows, then 56 of them were sold. How many cows are left?**

2) **The farm had 28 horses. Then, 4 more horses were born. How many horses are there now?**

3) **One barn has 73 pigs in it and another barn has 76 pigs in it. How many pigs are in the two barns combined?**

4) **There were 173 chickens on the farm. Then, the farmer bought 48 more. How many chickens are on the farm now?**

5) **In June, the chickens laid 1,704 eggs. In July, they laid 1,535 eggs. How many more eggs were laid in June than July?**

6) **The farmer brought 500 eggs to the market and sold 382 of them. How many eggs were not sold?**

7) **The farmer fed the chickens 78 bags of chicken feed last month and 67 bags of chicken feed this month. How many bags of chicken feed is that combined?**

8) **The farmer milked 235 cows in the morning and 226 cows in the evening. How many cows is that altogether?**

Name: Date: Score:

Unstoppable Owl

89) 1-Step Word Problems

1) Greg received 56 emails in the morning and 48 emails in the afternoon. How many emails did he receive?

2) Greg's first meeting lasted 38 minutes and his second meeting lasted 54 minutes. How many minutes is that altogether?

3) Greg is allowed to have a 45-minute lunch break. If his lunch break started 27 minutes ago, how much more time is he allowed for his lunch break?

4) Greg printed 462 copies of paper in the morning and 85 copies of paper in the afternoon. How many more copies of paper did he print in the morning than the afternoon?

5) There were 2,000 sheets of paper in the copy machine at the beginning of the week. Since then, 1,314 of the sheets have been used. How many sheets of paper are left in the machine?

6) Greg started working at the office 12 years ago. Jenny has been working at the office 15 more years than Greg has. How many years has Jenny been working at the office?

7) Greg needs to read a report that is 51 pages long. He's currently read 17 of the pages. How many more pages does he need to read?

8) It usually takes Greg 23 minutes to get home from work but took an extra 8 minutes today because of heavy traffic. How long did it take Greg to get home from work today?

Name: Date: Score:

Unstoppable Owl

90) 1-Step Word Problems

1) Tracy is going on vacation. Her flight is supposed to be 245 minutes long. If she's been on the flight for 128 minutes now, how much longer is the flight supposed to be?

2) Tracy took 78 pictures at the beach and 34 pictures in the city. How many pictures is that altogether?

3) Tracy's phone can store a total of 2,000 pictures. Her phone is already storing 1,281 pictures. How many more pictures can it store?

4) Tracy saw 29 monkeys and 42 parrots while on a hike. How many more parrots did she see than monkeys?

5) Hikers need to climb 306 steps to reach the scenic overlook. Tracy has already climbed 238 steps. How many more steps are left for her to climb?

6) It took Tracy 47 minutes to hike to the scenic overlook, but only took her 33 minutes to hike back to the beginning of the trail. How many minutes did she hike altogether?

7) The hotel allows a maximum of 120 people in the pool at a time. There are 94 people in the pool now. How many more people are allowed in the pool?

8) 190 chairs were set up for an outdoor concert. When the concert began, every chair had a person sitting on it, and there were 43 additional people standing. How many people were at the concert?

Name: Date: Score: *Unstoppable Owl*

91) 1-Step Word Problems

1) **Lisa is going to visit her grandchildren. It takes 72 minutes to drive to their house. Lisa has already been driving 46 minutes. How much longer is the drive expected to take?**

2) **Lisa brought with her 15 sugar cookies and 28 chocolate chip cookies. How many cookies is that altogether?**

3) **Lisa took the kids to the zoo. The zoo had 36 flamingoes and 63 penguins. How many more penguins were there than flamingoes?**

4) **The animal trainer said she feeds the orca 75 fish in the morning and 62 fish in the evening. How many fish is that altogether?**

5) **The trainer said the orca was 180 kg when it was born, and now it weighs 9,750 kg. How much heavier is the orca now than when it was born?**

6) **677 butterflies hatched last week, and 749 butterflies hatched this week. How many butterflies hatched in the two weeks combined.**

7) **The butterfly garden has 2,037 yellow flowers and 1,362 pink flowers. How many more yellow flowers are there than pink flowers?**

8) **Lisa's grandson scored 14 points in the first half of his basketball game and scored 13 points in the second half. What is the total number of points that he scored?**

Name: Date: Score: *Unstoppable Owl*

92) 1-Step Word Problems

1) Jake started a new daily workout routine. He lifts weights for 18 minutes then does cardio exercises for 35 minutes. How long is his daily workout routine altogether?

2) Jake was only able to do 12 pushups when he first started his routine. Now he can do 30 pushups. How many more pushups can Jake do now than when he first started the routine?

3) Jake's smartwatch said he had taken 4,582 steps the first half of the day and 8,952 steps the second half of the day. How many steps was that throughout the day?

4) Last week, Jake had 72,536 steps. This week, he had 90,704 steps. How many more steps did he have this week than last week.

5) Jake's resting heart rate is 72 beats per minute (bpm). When riding his bike, his heart rate was 131 bpm. How much faster was his heart beating while biking than resting?

6) Jake's doctor recommended that his heart rate not exceed 170 bpm while working out. How many bpm more is that than the 131 bpm while he was biking?

7) Today, Jake walked 35 minutes by himself and walked 27 minutes with a friend. What was the total amount of time that Jake spent walking today?

8) Jake kept a record of his workouts. After 100 days, there had only been 8 days when he did not complete the workout routine. How many days did he complete the workout routine?

Name: Date: Score: Unstoppable Owl

93) 1-Step Word Problems

1) Kyle studies dolphins. He saw two pods of dolphins this morning. One pod had 14 dolphins, and the other had 27 dolphins. How many dolphins is that combined?

2) The largest pod Kyle saw today had 27 dolphins. But he once saw a pod that had 114 dolphins. How many more dolphins were in that pod than the pod of 27 dolphins?

3) Last year, Kyle counted 154 dolphins living near the bay. This year, he has only counted 125 dolphins near the bay. How many fewer dolphins is that this year than last year?

4) Kyle placed 26 radio transmitters on dolphins last year and 19 radio transmitters on dolphins this year. How many transmitters is that altogether?

5) The transmitter from one dolphin showed that it swam 44 km in the morning and 35 km in the afternoon. How much further did it swim in the morning than in the afternoon?

6) The dolphin being tracked is 17 years old. Dolphins of this species often live to be 50 years old. How many more years will this dolphin need to live to reach the age of 50 years old?

7) Two dolphins are hunting together. One dolphin ate 34 squids, and the other dolphin ate 38 squids. How many squids did the two dolphins eat together?

8) The first dolphin stayed underwater for 482 seconds. The second dolphin stayed under water for 636 seconds. How much longer did the second dolphin stay under water than the first?

Name: Date: Score: *Unstoppable Owl*

94) 1-Step Word Problems

1) Clair is learning to paint. She bought 2 packs of paper to paint on. Each pack of paper has 148 sheets of paper. How many sheets of paper did she buy altogether?

2) Clair made 36 paintings last week and 45 paintings this week. How many paintings were made during both weeks combined?

3) So far, Clair has painted 155 flowers and 59 animals. How many more flowers has she painted than animals?

4) Clair brought 120 of her paintings to an art show to sell. When the show was over, 43 of the paintings had not sold. How many of the paints were sold?

5) Clair's friend asked if she could make a dragon painting for her. Clair painted the dragon for 75 minutes in the morning and 84 minutes in the evening. How much time did she spend painting the dragon?

6) The paint takes 90 minutes to dry. If Clair finished painting 23 minutes ago, how much more time will need to pass before the paint is dry?

7) Clair went to an art museum to get inspiration for her paintings. The museum has 1,507 paintings on the first floor and 1,724 paintings on the second floor. How many paintings is that altogether?

8) One of the paintings was bought at a yard sale for $145. After realizing it was a rare painting, it was sold to the museum for $340,000. How much more money did the museum pay for the painting than the person who bought it at the yard sale?

Name: Date: Score: Unstoppable Owl

95) 1-Step Word Problems

1) **Paul and Karen are going camping. They saw 37 deer the first day and 24 deer the second day. How many deer is that altogether?**

2) **Karen was 14 years old the last time she camped by the mountain. Now she is 52 years old. How many years have passed since she last camped there?**

3) **There are 118 campsites in the campground. Only 35 of the sites are not being used. How many campsites are being used?**

4) **It took them 76 minutes to hike up the mountain, but only 45 minutes to hike back down. How long did the hike take altogether?**

5) **Paul and Karen packed 30 marshmallows for the trip. They've used 13 of them to make smores. How many marshmallows are left?**

6) **They stayed up late by the campfire and watched a meteor shower. Paul saw 27 shooting stars and Karen saw 34. How many more shooting stars did Karen see than Paul?**

7) **Paul and Karen were watching a flock of 65 swans. Then, another flock with 46 swans joined the first flock. How many swans is that altogether?**

8) **A tree at the campsite is believed to be 237 years old. A park ranger said the tree may live to be 400 years old. How many more years need to pass for the tree to reach the age of 400 years?**

Name: Date: Score: *Unstoppable Owl*

96) 2-Step Word Problems

Some word problems require more than one step to find the solution. The problems in this section will be 2-step word problems. Instead of just adding or subtracting to find the solution, you may first need to add or subtract, and then add or subtract again.

Example 1: (adding then subtracting)

A pet shop had 58 goldfish. Then, 36 more goldfish were born, and 42 goldfish were sold. How many goldfish does the pet shop have now?

Step 1: Add

```
  1
  58
 +36
 ---
  94
```

54 + 36 = **94**

Step 2: Subtract

```
  94
 -42
 ---
  52
```

94 – 42 = **52**

Example 2: (adding then adding)

The pet shop had 14 snakes. Then the pet shop bought 9 more snakes and 34 baby snakes were born. How many snakes does the pet shop have now?

Step 1: Add

```
  1
  14
 + 9
 ---
  23
```

14 + 9 = **23**

Step 2: Subtract

```
  23
 +34
 ---
  57
```

23 + 34 = **57**

1) **Amy is a teacher. There were 14 girls and 17 boys in the class. Then, 2 more girls joined the class after the school year started. How many students are in her class now?**

2) **Amy's class had 520 crayons but 185 of them were thrown away once they were too used or broken. Amy then bought 240 more crayons for the class. How many crayons does the class have now?**

3) **The school had 2,500 pencils. Then, 1,652 pencils were given to the older students, and 793 pencils were given to the younger students. How many pencils does the school have left?**

4) **The school serves 164 students during the first lunch period, 183 students during the second lunch period, and 155 students during the third lunch period. How many students are served lunch altogether?**

Name: Date: Score: *Unstoppable Owl*

97) 2-Step Word Problems

1) Scott's company builds houses. The company had 300,000 nails. Then, 94,315 nails were used in June and 75,846 were used in July. How many nails does his company have left?

2) Scott's company used 24,905 screws in June, 27,146 screws in July, and 23,083 screws in August. How many screws were used in these three months combined?

3) Scott estimated that it would take 39 days to build a house. But an additional 14 days were needed to complete the basement and an additional 8 days were needed for the roof. How many days did it take to build this house?

4) The company had 12,000 boards. 8,713 boards were used in June, so Scott's company purchased 6,250 more boards. How many boards does the company have now?

5) Scott needs to transport the 6,250 boards. His trailer transported 2,736 boards in the first load and 2,665 boards in the second load. How many more boards still need to be transported?

6) Scott needs to pay his crew for their working hours. If they worked 427 hours this week, 532 hours last week, and 476 hours the week before that, how many hours did they work during these three weeks combined?

7) It was supposed to take the cement truck 37 minutes to drive to the construction site, but the driver was 15 minutes late. After arriving, it took 52 minutes to pour the cement. How much time is that altogether?

8) Scott's company applied for 54 building permits. 28 of the permits were approved last week, and 15 were approved this week. How many more permits have not yet been approved?

Name: Date: Score:

Unstoppable Owl

98) 2-Step Word Problems

1) Eric's orchard has 538 apple trees, 164 pear trees, and 745 cherry trees. How many trees is that altogether?

2) 932 customers came to the orchard in July, 1,485 in August, and 4,804 in September. How many more customers came in September than July and August combined?

3) 6,723 apples were harvested this year. 2,582 apples were sold the first week after the harvest and 1,759 were sold the second week after. How many apples have not been sold yet?

4) 875 containers of cherries were picked this week, 903 containers of cherries were picked last week, and 764 containers were picked the week before that. How many containers of cherries is that altogether?

5) The orchard had 538 apple trees. Then, 73 of them blew over during a storm. After the storm, Eric planted 125 more apple trees. How many apple trees are there now?

6) Eric had 500 boxes for customers to put their fruit in. Customers used 286 boxes this week, so Eric bought 250 more boxes. How many boxes does Eric have now?

7) Eric made a commercial to advertise the orchard. He paid a radio station to play the commercial 100 times. The station played it 45 times the first week and 28 times the second week. How many more times should the commercial be played?

8) Eric brought 2,000 of his pears to a farmers' market. On Friday, he sold 548 pears and on Saturday, he sold 1,215 pears. How many pears have not been sold yet?

Name: Date: Score: *Unstoppable Owl*

99) 2-Step Word Problems

1) Jodi's bakery had 78 pies. Then, 19 were sold in the morning and 53 were sold in the afternoon. How many pies are left?

2) Jodi baked 421 cupcakes on Monday, 385 cupcakes on Tuesday, and 407 cupcakes on Wednesday. How many cupcakes is that altogether?

3) Of the 421 cupcakes baked on Monday, 235 were vanilla cupcakes and the rest were chocolate chunk cupcakes. How many were chocolate chunk cupcakes?

4) Jodi had 400 cookies at the beginning of the day. 283 cookies were sold in the morning, so Jodi baked 140 more. How many cookies does she have now?

5) Of the 140 cookies baked, 45 of them had chocolate chips, 36 had raisins, and the rest were sugar cookies. How many were sugar cookies?

6) Jodi had 23,000 napkins. Then, her customers used 1,794 napkins one week and 2,238 napkins the next week. How many napkins does Jodi have left?

7) Jodi baked 137 blueberry muffins, 86 cranberry muffins, and 59 banana nut muffins. What is the total number of muffins Jodi baked?

8) Jodi sold 106 blueberry muffins, 86 cranberry muffins, and 44 banana nut muffins. How many muffins did she sell altogether?

Name: Date: Score: *Unstoppable Owl*

100) 2-Step Word Problems

1) **A farm had 1,054 chickens. Then, 295 of them were sold in March and 158 were sold in April. How many chickens does the farm have left?**

2) **The chickens laid 1,249 eggs in June, 1,064 eggs in July, and 1,176 eggs in August. How many eggs is that during these three months combined?**

3) **The farm had 329 pigs. Then, 85 piglets were born and 68 of the adult pigs were sold. How many pigs does the farm have now?**

4) **The farm had 105 sacks of oats. Then, 47 of the sacks were fed to the animals and the farmer purchased 72 more sacks. How many sacks of oats does the farm have now?**

5) **The farmer plowed 136 rows today, 145 rows yesterday, and 128 rows the day before that. What is the total number of rows that were plowed?**

6) **The farmer had 50,000 seeds. Then, 18,539 of the seeds were planted on Monday and 23,617 seeds were planted on Tuesday. How many seeds does the farmer have left?**

7) **The farmer had 452 cows and 267 sheep. Then, the farmer bought 89 more sheep. Now, how many more cows does the farmer have than sheep?**

8) **The farmer had 314 bales of hay. Then, 75 bales were fed to the cows and 56 bales were fed to the sheep. How many bales of hay are left?**

101) Answer Key

Page 1:
1) 68 2) 95 3) 79 4) 67
5) 78 6) 26 7) 35 8) 89

Page 2:
1) 74 2) 78 3) 58 4) 94
5) 75 6) 69 7) 28 8) 86
9) 46 10) 79 11) 68 12) 46
13) 55 14) 85 15) 67 16) 51
17) 59 18) 70 19) 39 20) 53
21) 89 22) 72 23) 66 24) 99

Page 3:
1) 28 2) 86 3) 79 4) 44
5) 97 6) 67 7) 16 8) 28
9) 58 10) 18 11) 38 12) 49

Page 4:
1) 77 2) 85 3) 68 4) 48
5) 48 6) 55 7) 86 8) 97
9) 48 10) 67 11) 82 12) 96
13) 57 14) 60 15) 79 16) 45
17) 68 18) 77 19) 89 20) 33
21) 98 22) 47 23) 40 24) 85

Page 5:
1) 62 2) 94 3) 69 4) 65
5) 45 6) 80 7) 91 8) 88

Page 6:
1) 124 2) 108 3) 180 4) 103
5) 97 6) 93 7) 107 8) 171

Page 7:
1) 128 2) 102 3) 83 4) 131
5) 109 6) 110 7) 77 8) 194
9) 141 10) 112 11) 90 12) 101
13) 152 14) 90 15) 119 16) 115
17) 65 18) 108 19) 143 20) 68
21) 156 22) 101 23) 106 24) 121

Page 8:
1) 128 2) 105 3) 81 4) 141
5) 145 6) 118 7) 109 8) 140
9) 103 10) 77 11) 127 12) 118
13) 106 14) 153 15) 120 16) 122
17) 121 18) 88 19) 162 20) 80
21) 168 22) 100 23) 170 24) 124

Page 9:
1) 123 2) 93 3) 159 4) 116
5) 106 6) 80 7) 126 8) 143
9) 110 10) 145 11) 131 12) 171
13) 172 14) 121 15) 122 16) 150
17) 88 18) 103 19) 111 20) 134
21) 111 22) 173 23) 102 24) 150

Page 10:
1) 143 2) 116 3) 94 4) 129
5) 98 6) 168 7) 121 8) 108
9) 198 10) 140 11) 122 12) 132
13) 64 14) 116 15) 128 16) 141
17) 93 18) 180 19) 125 20) 120
21) 122 22) 82 23) 79 24) 164

Page 11:
1) 83 2) 92 3) 117 4) 170
5) 41 6) 137 7) 114 8) 96

Page 12:
1) 72 2) 141 3) 191 4) 151
5) 81 6) 140 7) 112 8) 70
9) 105 10) 54 11) 151 12) 92
13) 90 14) 171 15) 116 16) 67
17) 171 18) 85 19) 140 20) 102

Page 13:
1) 122 2) 53 3) 63 4) 148
5) 101 6) 162 7) 101 8) 148
9) 77 10) 119 11) 108 12) 183
13) 135 14) 152 15) 150 16) 88
17) 172 18) 85 19) 80 20) 134

Page 14:
1) 152 2) 172 3) 116 4) 140
5) 148 6) 72 7) 130 8) 34
9) 161 10) 192 11) 91 12) 33
13) 52 14) 81 15) 131 16) 107
17) 93 18) 81 19) 170 20) 121

Page 15:
1) 1,232 2) 1,549 3) 846
4) 1,600 5) 898 6) 1,502
7) 1,227 8) 839 9) 1,029

Page 16:
1) 550 2) 755 3) 1,420
4) 1,681 5) 922 6) 336
7) 1,013 8) 1,418 9) 1,205
10) 1,001 11) 737 12) 1,514
13) 1,145 14) 1,010 15) 1,998
16) 800 17) 995 18) 1,332

Page 17:
1) 1,454 2) 1,202 3) 1,286
4) 919 5) 1,440 6) 1,007
7) 1,223 8) 574 9) 1,021
10) 402 11) 1,502 12) 825
13) 1,085 14) 1,918 15) 1,513
16) 762 17) 1,510 18) 1,020

Page 18:
1) 439 2) 1,130 3) 1,301
4) 737 5) 670 6) 1,683
7) 1,303 8) 1,130 9) 1,250
10) 1,071 11) 631 12) 1,010
13) 1,333 14) 769 15) 593
16) 1,290 17) 463 18) 1,087

Page 19:
1) 1,021 2) 1,339 3) 1,201
4) 1,554 5) 1,521 6) 1,250
7) 865 8) 1,261 9) 1,018
10) 1,021 11) 1,343 12) 960
13) 1,231 14) 1,047 15) 508
16) 1,060 17) 901 18) 1,686

Page 20:
1) 1,557 2) 1,010 3) 1,006
4) 225 5) 1,220 6) 1,201
7) 1,724 8) 689 9) 1,005
10) 364 11) 1,930 12) 1,230

Page 21:
1) 844 2) 768 3) 1,812
4) 1,011 5) 350 6) 1,202
7) 1,032 8) 768 9) 1,641
10) 956 11) 434 12) 1,114
13) 1,030 14) 1,812 15) 1,342

Page 22:
1) 1,300 2) 539 3) 1,240
4) 542 5) 881 6) 1,001
7) 1,221 8) 841 9) 323
10) 1,128 11) 358 12) 1,211
13) 775 14) 1,279 15) 437

Page 23:
1) 852 2) 760 3) 796
4) 1,881 5) 1,098 6) 1,201
7) 1,620 8) 818 9) 1,576
10) 1,175 11) 1,168 12) 767
13) 1,100 14) 1,813 15) 1,530

Page 24:
1) 1,352 2) 805 3) 661
4) 610 5) 1,694 6) 1,046
7) 1,001 8) 1,044 9) 1,926
10) 510 11) 791 12) 1,212
13) 1,425 14) 645 15) 1,704

Page 25:
1) 12,270 2) 13,971
3) 92,167 4) 186,424
5) 1,036,424 6) 1,242,303

Page 26:
1) 7,859 2) 12,532
3) 2,584 4) 13,232
5) 90,258 6) 100,172
7) 95,752 8) 184,507
9) 1,097,503 10) 1,333,371
11) 517,147 12) 1,003,295

Page 27:
1) 9,154 2) 15,818
3) 16,990 4) 12,442
5) 170,730 6) 74248
7) 114,932 8) 64,415
9) 435,854 10) 1,465,229
11) 1,529,301 12) 872,340

Page 28:
1) 10,349 2) 12,550
3) 7,345 4) 18,048
5) 80,065 6) 123,330
7) 115,852 8) 82,650
9) 609,031 10) 1,371,016
11) 1,987,654 12) 1,144,983

Page 29:
1) 9,559 2) 12,901
3) 6,366 4) 18,040
5) 114,385 6) 98,623
7) 34,749 8) 109,141
9) 506,163 10) 1,530,133
11) 1,014,068 12) 811,637

Page 30:
1) 9,092 2) 7,836
3) 7,136 4) 12,032
5) 142,439 6) 66,248
7) 181,129 8) 124,130
9) 787,084 10) 1,341,659

Page 31:
1) 6,257 2) 8,581
3) 12,561 4) 10,296
5) 125,781 6) 74,328
7) 81,353 8) 112,339
9) 1,278,196 10) 855,094

Page 32:
1) 9,175 2) 3,353
3) 7,123 4) 3,764
5) 159,441 6) 61,870
7) 33,363 8) 55,247
9) 572,063 10) 891,617

Page 33:
1) 10,480 2) 3,222
3) 8,102 4) 16,975
5) 63,804 6) 117,526
7) 57,238 8) 51,472
9) 1,128,041 10) 966,272

Page 34:
1) 11,665 2) 12,782
3) 6,351 4) 7,471
5) 31,032 6) 42,613
7) 114,150 8) 117,851
9) 530,283 10) 548,751

Page 35:
1) 51 2) 30 3) 4 4) 84
5) 21 6) 11 7) 43 8) 18
9) 16 10) 1 11) 18 12) 20

Page 36:
1) 81 2) 32 3) 24 4) 47
5) 4 6) 55 7) 16 8) 32
9) 64 10) 31 11) 7 12) 22
13) 71 14) 33 15) 31 16) 40
17) 76 18) 23 19) 34 20) 30
21) 51 22) 50 23) 7 24) 12

Page 37:
1) 35 2) 50 3) 52 4) 41
5) 44 6) 30 7) 11 8) 28
9) 60 10) 84 11) 12 12) 42
13) 41 14) 63 15) 5 16) 25
17) 33 18) 13 19) 10 20) 26
21) 23 22) 21 23) 21 24) 68

Page 38:
1) 17 2) 25 3) 45 4) 22
5) 14 6) 62 7) 40 8) 42
9) 52 10) 35 11) 12 12) 71
13) 42 14) 47 15) 35 16) 14
17) 12 18) 50 19) 41 20) 15
21) 30 22) 74 23) 43 24) 41

Page 39:
1) 33 2) 83 3) 32 4) 17
5) 13 6) 18 7) 15 8) 24
9) 52 10) 32 11) 44 12) 26
13) 80 14) 16 15) 21 16) 44
17) 23 18) 72 19) 26 20) 5
21) 50 22) 11 23) 35 24) 22

Page 40:
1) 45 2) 39 3) 24 4) 68
5) 14 6) 29 7) 29 8) 45

Page 41:
1) 18 2) 19 3) 8 4) 58
5) 17 6) 78 7) 25 8) 4
9) 48 10) 19 11) 28 12) 37
13) 15 14) 37 15) 28 16) 46
17) 49 18) 19 19) 28 20) 42
21) 27 22) 16 23) 25 24) 48

Page 42:
1) 27 2) 43 3) 25 4) 28
5) 23 6) 47 7) 9 8) 26
9) 23 10) 24 11) 36 12) 13
13) 39 14) 47 15) 18 16) 54
17) 38 18) 28 19) 44 20) 19
21) 18 22) 29 23) 36 24) 57

Page 43:
1) 27 2) 42 3) 35 4) 27
5) 18 6) 37 7) 15 8) 18
9) 35 10) 57 11) 14 12) 18
13) 47 14) 27 15) 39 16) 32
17) 37 18) 8 19) 14 20) 18
21) 43 22) 14 23) 58 24) 19

Page 44:
1) 28 2) 7 3) 42 4) 29
5) 18 6) 17 7) 66 8) 18
9) 56 10) 17 11) 58 12) 24
13) 4 14) 13 15) 21 16) 7
17) 77 18) 26 19) 44 20) 18
21) 38 22) 26 23) 38 24) 7

Page 45:
1) 28 2) 35 3) 29 4) 19
5) 37 6) 26 7) 18 8) 36
9) 37 10) 49 11) 18 12) 26
13) 20 14) 45 15) 28 16) 29

Page 46:
1) 48 2) 9 3) 24 4) 29
5) 53 6) 64 7) 57 8) 27
9) 38 10) 16 11) 27 12) 17
13) 38 14) 19 15) 35 16) 50
17) 65 18) 18 19) 48 20) 27

Page 47:
1) 34 2) 29 3) 45 4) 47
5) 29 6) 27 7) 17 8) 49
9) 21 10) 46 11) 18 12) 27
13) 8 14) 16 15) 64 16) 18
17) 18 18) 28 19) 69 20) 38

Page 48:
1) 32 2) 13 3) 48 4) 28
5) 25 6) 32 7) 76 8) 49
9) 29 10) 56 11) 18 12) 27
13) 24 14) 28 15) 37 16) 24
17) 4 18) 18 19) 2 20) 25

Page 49:
1) 29 2) 36 3) 46 4) 78
5) 8 6) 36 7) 65 8) 18
9) 39 10) 32 11) 35 12) 28
13) 34 14) 79 15) 29 16) 49
17) 40 18) 7 19) 27 20) 47

Page 50:
1) 84 2) 434 3) 56
4) 597 5) 349 6) 329
7) 574 8) 174 9) 239

Page 51:
1) 565 2) 377 3) 179
4) 504 5) 495 6) 72
7) 374 8) 583 9) 71
10) 289 11) 204 12) 264
13) 95 14) 187 15) 275
16) 588 17) 399 18) 114

Page 52:
1) 279 2) 158 3) 447
4) 28 5) 506 6) 89

Page 53:
1) 368 2) 368 3) 357
4) 284 5) 268 6) 279
7) 304 8) 36 9) 158
10) 146 11) 218 12) 387
13) 404 14) 278 15) 379
16) 366 17) 191 18) 664

102) Answer Key

Page 54:
1) 496 2) 723 3) 168
4) 194 5) 361 6) 169
7) 227 8) 475 9) 182
10) 259 11) 36 12) 584
13) 399 14) 594 15) 74
16) 327 17) 167 18) 299

Page 55:
1) 83 2) 279 3) 438
4) 62 5) 178 6) 439
7) 178 8) 148 9) 413
10) 290 11) 189 12) 642

Page 56:
1) 264 2) 73 3) 163
4) 357 5) 684 6) 380
7) 178 8) 152 9) 53
10) 38 11) 354 12) 273
13) 377 14) 25 15) 26

Page 57:
1) 229 2) 387 3) 475
4) 324 5) 486 6) 299
7) 493 8) 397 9) 119
10) 213 11) 199 12) 267
13) 566 14) 442 15) 389

Page 58:
1) 84 2) 282 3) 258
4) 249 5) 493 6) 77
7) 70 8) 309 9) 78
10) 173 11) 188 12) 347
13) 666 14) 428 15) 128

Page 59:
1) 283 2) 239 3) 649
4) 164 5) 736 6) 331
7) 557 8) 265 9) 17
10) 553 11) 228 12) 229
13) 120 14) 383 15) 383

Page 60:
1) 1,938 2) 445
3) 37,608 4) 36,729
5) 392,757 6) 80,062

Page 61:
1) 2,684 2) 1,059
3) 2,952 4) 2,167
5) 399 6) 51,830
7) 14,665 8) 16,219
9) 198,180 10) 286,764
11) 372,911 12) 648,837

Page 62:
1) 438 2) 6,087
3) 618 4) 6,482
5) 27,866 6) 34,569
7) 25,828 8) 29,817
9) 60,735 10) 273,899
11) 317,786 12) 163,723

Page 63:
1) 4,772 2) 1,229
3) 3,668 4) 5,534
5) 7,592 6) 7,729
7) 59,352 8) 27,915
9) 68,446 10) 202,790
11) 484,619 12) 65,633

Page 64:
1) 3,806 2) 3,684
3) 5,863 4) 4,657
5) 19,778 6) 19,549
7) 42,310 8) 15,918
9) 280,977 10) 638,729
11) 388,221 12) 66,775

Page 65:
1) 4,291 2) 5,959
3) 4,567 4) 2,796
5) 27,473 6) 18,523
7) 338,571 8) 179,367

Page 66:
1) 1,228 2) 5,558
3) 1,620 4) 4,914
5) 56,762 6) 9,558
7) 21,896 8) 63,570
9) 417,765 10) 152,746

Page 67:
1) 245 2) 4,833
3) 3,165 4) 774
5) 29,748 6) 28,913
7) 26,356 8) 3,611
9) 294,312 10) 306,705

Page 68:
1) 4,829 2) 2,179
3) 839 4) 1,388
5) 1,675 6) 29,197
7) 60,513 8) 85,576
9) 175,078 10) 451,075

Page 69:
1) 3,691 2) 3,919
3) 6,635 4) 738
5) 26,663 6) 6,827
7) 37,228 8) 16,022
9) 354,571 10) 145,308

Page 70:
1) 131 2) 177 3) 84 4) 114
5) 14,001 6) 83,329 7) 116,820
8) 37 9) 15 10) 48 11) 17
12) 3,495 13) 18,605 14) 28,785

Page 71:
1) 112 2) 150 3) 142 4) 93
5) 11,633 6) 123,181 7) 182,582
8) 31 9) 34 10) 19 11) 35
12) 3,319 13) 38,835 14) 47,662

Page 72:
1) 124 2) 102 3) 161 4) 42
5) 5,633 6) 103,119 7) 42,792
8) 9 9) 65 10) 39 11) 25
12) 6,789 13) 16,559 14) 59,575

Page 73:
1) 123 2) 179 3) 122 4) 82
5) 13,497 6) 124,339 7) 30,255
8) 32 9) 23 10) 20 11) 38
12) 2,371 13) 32,737 14) 8,694

Page 74:
1) 94 2) 136 3) 130 4) 122
5) 11,809 6) 126,516 7) 53,234
8) 19 9) 29 10) 67 11) 61
12) 1,259 13) 19,429 14) 34,973

Page 75:
1) 45 2) 92 3) 131 4) 110
5) 4,442 6) 112,870 7) 51,534
8) 25 9) 26 10) 18 11) 27
12) 2,684 13) 17,479 14) 53,684

Page 76:
1) 154 2) 154 3) 111 4) 133
5) 14,027 6) 86,739 7) 175,320
8) 37 9) 47 10) 49 11) 16
12) 803 13) 42,774 14) 8,162

Page 77:
1) 142 2) 196 3) 61 4) 142
5) 10,301 6) 42,326 7) 72,241
8) 37 9) 47 10) 17 11) 24
12) 5,979 13) 17,575 14) 8,847

Page 78:
1) 111 2) 147 3) 70 4) 85
5) 5,104 6) 106,217 7) 47,439
8) 16 9) 24 10) 18 11) 39
12) 677 13) 9,658 14) 37,228

Page 79:
1) 102 2) 112 3) 126 4) 68
5) 12,158 6) 111,864 7) 160,330
8) 16 9) 18 10) 39 11) 62
12) 1,785 13) 58,780 14) 849

Page 80:
1) 162 2) 61 3) 132 4) 131
5) 13,800 6) 132,581 7) 81,840
8) 28 9) 29 10) 22 11) 49
12) 59 13) 33,525 14) 52,874

Page 81:
1) 177 2) 121 3) 58 4) 153
5) 10,316 6) 68,697 7) 114,232
8) 31 9) 46 10) 19 11) 8
12) 1,181 13) 7,829 14) 37,548

Page 82:
1) 125 2) 123 3) 133 4) 147
5) 3,502 6) 124,903 7) 44,992
8) 27 9) 26 10) 40 11) 48
12) 3,689 13) 19,547 14) 16,709

Page 83:
1) 123 2) 79 3) 181 4) 123
5) 5,533 6) 43,218 7) 171,339
8) 19 9) 47 10) 16 11) 59
12) 405 13) 34,684 14) 21,887

Page 84:
1) 92 2) 99 3) 73 4) 146
5) 15,306 6) 133,777 7) 54,546
8) 18 9) 13 10) 29 11) 78
12) 8,726 13) 36,479 14) 4,276

Page 85:
1) 120 2) 142 3) 107 4) 158
5) 11,940 6) 156,628 7) 46,055
8) 9 9) 14 10) 66 11) 37
12) 3,867 13) 48,169 14) 14,787

Page 86:
1) 103 seeds 2) 23 more
3) 76 carrots 4) 80 tomatoes
5) 431 potatoes 6) 83 potatoes

Page 87:
1) 846 cups 2) 49 more
3) 264 bags 4) 322 bags
5) 36 chairs 6) 57 chairs
7) 203 customers 8) 231 customers

Page 88:
1) 256 cows 2) 32 horses
3) 149 pigs 4) 221 chickens
5) 169 eggs 6) 118 eggs
7) 145 bags 8) 461 cows

Page 89:
1) 104 emails 2) 92 minutes
3) 18 minutes 4) 377 copies
5) 686 sheets 6) 27 years
7) 34 pages 8) 31 minutes

Page 90:
1) 117 minutes 2) 112 pictures
3) 719 pictures 4) 13 more
5) 68 steps 6) 80 minutes
7) 26 more 8) 233 people

Page 91:
1) 26 minutes 2) 43 cookies
3) 27 more 4) 137 fish
5) 9,570 kg 6) 1,426 butterflies
7) 675 more 8) 27 points

Page 92:
1) 53 minutes 2) 18 more
3) 13,534 steps 4) 18,168 more
5) 59 bpm 6) 39 bpm
7) 62 minutes 8) 92 days

Page 93:
1) 41 dolphins 2) 87 more
3) 29 fewer 4) 45 transmitters
5) 9 km 6) 33 years
7) 72 squids 8) 154 seconds

Page 94:
1) 296 sheets 2) 81 paintings
3) 96 more 4) 77 paintings
5) 159 minutes 6) 67 minutes
7) 3,231 paintings 8) $339,855 more

Page 95:
1) 61 deer 2) 38 years
3) 83 campsites 4) 121 minutes
5) 17 marshmallows 6) 7 more
7) 111 swans 8) 163 years

Page 96:
1) 33 students 2) 575 crayons
3) 55 pencils 4) 502 students

Page 97:
1) 129,839 nails 2) 75,134 screws
3) 61 days 4) 9,537 boards
5) 849 boards 6) 1,435 hours
7) 104 minutes 8) 11 permits

Page 98:
1) 1,447 trees 2) 2,387 more
3) 2,382 apples 4) 2,542 containers
5) 590 apple trees 6) 464 boxes
7) 27 more 8) 237 pears

Page 99:
1) 6 pies 2) 1,213 cupcakes
3) 186 chocolate chunk 4) 257 cookies
5) 59 sugar cookies 6) 18,968 napkins
7) 282 muffins 8) 236 muffins

Page 100:
1) 601 chickens 2) 3,489 eggs
3) 346 pigs 4) 130 sacks
5) 409 rows 6) 7,844 seeds
7) 96 more 8) 183 bales

ISBN: 978-1-63578-551-7
Current contact information can be found at:
www.UnstoppableOwl.com
www.LibroStudioLLC.com

Disclaimers:
This book and its contents are provided "AS IS" without warranty of any kind, expressed or implied, and hereby disclaims all implied warranties, including any warranty of merchantability and warranty of fitness for a particular purpose.

The creator and publisher DO NOT GUARANTEE THE ACCURACY, RELIABILITY, OR COMPLETENESS OF THE CONTENT OF THIS BOOK OR RELATED RESOURSES AND IS NOT RESPONSIBLE FOR ANY ERRORS OR OMISSIONS. We apologize for any inaccurate, outdated, or misleading information. Feel free to contact us if you have questions or concerns.

ALWAYS SEEK ADVICE FROM A QUALIFIED PROFESSIONAL BEFORE USING THIS BOOK OR MAKING DECISIONS BASED ON THE INFORMATION FOUND IN THIS BOOK OR RELATED RESOURCES. The creator and publisher are not liable for any decision made or action taken based on this book's content and information nor that of any related resource. You and any other persons are responsible for your own judgments, decisions, and actions.

Libro Studio LLC publishes books and other content in a variety of print and electronic formats. Some content that appears in one format may not be available in other formats. For example, some content found in a print book may not be available in the eBook format, and vice versa. Furthermore, Libro Studio LLC reserves the right to update, alter, unpublish, and/or republish the content of any of these formats at any time.

Made in United States
North Haven, CT
05 November 2026

21771729R10059